MATERIALS SCIENCE AND TECHNOLOGIES

RECENT STUDIES IN MATERIALS SCIENCE

MATERIALS SCIENCE AND TECHNOLOGIES

Additional books and e-books in this series can be found on Nova's website under the Series tab.

MATERIALS SCIENCE AND TECHNOLOGIES

RECENT STUDIES IN MATERIALS SCIENCE

PATRICK R. LIND
EDITOR

nova
science publishers
New York

Library of Congress Cataloging-in-Publication Data

ISBN: 978-1-53615-270-8

Published by Nova Science Publishers, Inc. † New York

CONTENTS

PREFACE

The goal of the first chapter is to express the main achievements in use of the Jahn-Teller effect (JTE) and pseudo JTE (PJTE) in search of materials with enhanced or novel properties, are presented as a part of what is known as materials science.

The hallmark of this entire chapter is the commitment to identify situations in which "waves" are produced and provide a systematic classification of such convective instabilities together with a description of related features based on advanced numerical simulations.

The quality and mechanical or electrical properties of the resulting solids and crystals are adversely affected by thermogravitational convection as it can induce defects in their atomic or molecular structure (this is the case, e.g., of typical crystal-growth techniques such as the horizontal Bridgman (HB), the Floating zone (FZ) or the Czochralski (CZ) methods). This present chapter aims to present a focused review of landmark (past) and very recent contributions on the nature, structure and hierarchy of instabilities of this type of convection.

This chapter is devoted to the investigation of the thermo-kinetics features of overcooled austenite transformation in three grades of 0.75-0.77 wt.% C complex-alloyed (Mn-Si, Mn-Si-Cr-Mo) steel intended for rail applications. Magnetometric and thermal analyses were used to build Time-Temperature-Transformation and Continuous-Cooling-Transformation diagrams presenting the kinetics of austenite transformation in studied steels.

Chapter 1 - In a review chapter, including also original content, the main achievements in use of the Jahn-Teller effect (JTE) and pseudo JTE (PJTE) in search of materials with enhanced or novel properties, are presented as a part of what is known as materials science. First, the authors introduce briefly the physics beyond these first-principle effects, and their main features, important in applications. A specific outcome from these effects, the *dynamics of the local distortions* in molecular systems and local centers in solids triggers novel properties and/or enhances some existing properties by orders of magnitudes. On a larger scale, the PJTE induced local dipolar dynamics, due to intercell interactions, results in ferroelectric properties of perovskite crystals with temperature-dependent order-disorder phase transitions and a variety of special effects, confirmed experimentally. Among them is the role of the spin in spontaneous polarization, explaining the origin and predicting multiferroic properties and magneto-electric effects. The theory elucidates also the origin of very strong flexoelectricity, permittivity, and electrostriction, enhanced by three-four orders of magnitude due to the PJTE dynamics, and explains the origin of polar nanoregions and relaxor properties of ferroelectrics. Among other features of the JTE and PJTE, important to materials science, the authors discuss the prediction of a whole class of molecular and solid state compounds with magnetic-dielectric bistability, illustrated by the example of the CuF_3 molecule, and $LiCuO_2$ and $NaCuO_2$ crystals. Somewhat different looks the application of the PJTE to manipulate the structure and properties of two-dimensional (2D) systems. Here the knowledge of the origin of their buckling as due to the PJTE is used to influence the latter by coordination to other ions or molecular groups, or by redox processes, or by targeted substitutions, all intended to control the buckling, important in applications. Successful realization of this PJTE methodology is demonstrated on a series of 2D and quasi-2D systems.

Chapter 2 - Zone melting (or zone refining or floating zone process, FZ) is a group of similar methods, specifically conceived for the purification of crystals, in which thermally-driven flows of both gravitational and surface-tension natures are typically produced when the considered material is processed. Since the melt never comes into contact with anything but vacuum (or inert gases), there are no contaminants that the melt may incorporate. Even though compounds with higher purity and improved quality can be obtained with this technique, a typical drawback is represented by the defects potentially induced in the crystalline structure by the unavoidable convection emerging in the fluid phase. In the present chapter, special attention is paid to a specific category of materials known as transparent oxides. A range of conditions is explored, differing in the dominant effect (buoyancy or Marangoni flow), the thermal conditions (heating being provided along the radial or axial direction) and the relative direction of gravity and applied temperature gradient. The hallmark of the entire chapter is the authors' commitment to identify situations in which "waves" are produced and provide a systematic classification of such convective instabilities together with a description of related features based on advanced numerical simulations.

Chapter 3 - Non isothermal flows of liquid metals induced by buoyancy are central to many advanced technological applications in materials science, often at the cutting-edge of modern engineering. They have indeed a significant impact on the production of many materials obtained via the solidification of a melt. The quality and mechanical or electrical properties of the resulting solids and crystals are adversely affected by thermogravitational convection as it can induce defects in their atomic or molecular structure (this is the case, e.g., of typical crystal-growth techniques such as the horizontal Bridgman (HB), the Floating zone (FZ) or the Czochralski (CZ) methods). The present chapter aims to present a focused review of landmark (past) and very recent contributions on the nature, structure and hierarchy of instabilities of this type of convection. In particular, starting from simple situations corresponding to steady and laminar flows and moving towards fully developed turbulence, the authors present the typical hydrodynamic and hydrothermal disturbances emerging in differentially heated liquid metals and clarify the relationship among their properties and general influential factors such as: the degree of confinement (aspect ratio), morphology (wall orientation in space) and spatial degrees of freedom (number of active dimensions) of the domain hosting the melt. Manifestations of these modes of convection (including, but not limited to, transverse waves travelling in the downstream or in the upstream direction, standing waves, modulated pulso-traveling disturbances, longitudinal waves and multi-wave patterns) are discussed in detail. More complex situations are placed in the context of existing theories on turbulence in fluids and treated using concepts, methods and tools typical of the chaotic systems analysis.

Chapter 4 - The present paper is devoted to the investigation of the thermo-kinetics features of overcooled austenite transformation in three grades of 0.75-0.77 wt.% C complex-alloyed (Mn-Si, Mn-Si-Cr-Mo) steel intended for rail applications. Magnetometric and thermal analyses were used to build Time-Temperature-Transformation and Continuous-Cooling-Transformation diagrams presenting the kinetics of austenite transformation in studied steels. Microstructural (LOM, SEM, TEM) techniques, XRD and microhardness measurements were employed to characterize the microstructure and the mechanisms of its formation. The cooling rate values for different microstructure type are determined for each studied steel. Different morphological types of bainite formed in steels depending on the transformation temperature (200 – 450°C) are observed and classified. The features of bainite fine substructure are described and discussed. The effects of the cooling rate and alloying elements on transformation characteristics are evaluated. The kinetics of austenite transformation in isothermal conditions is analyzed basing on Johnson–Mehl–Avrami equation and Cahn's theory.

In: Recent Studies in Materials Science ISBN: 978-1-53615-270-8
Editor: Patrick R. Lind

Chapter 1

THE JAHN-TELLER AND PSEUDO JAHN-TELLER EFFECTS IN MATERIALS SCIENCE

I. B. Bersuker[*]
Institute for theoretical chemistry,
The University of Texas at Austin, Austin, TX, US

ABSTRACT

In a review chapter, including also original content, the main achievements in use of the Jahn-Teller effect (JTE) and pseudo JTE (PJTE) in search of materials with enhanced or novel properties, are presented as a part of what is known as materials science. First, we introduce briefly the physics beyond these first-principle effects, and their main features, important in applications. A specific outcome from these effects, the *dynamics of the local distortions* in molecular systems and local centers in solids triggers novel properties and/or enhances some existing properties by orders of magnitudes. On a larger scale, the PJTE induced local dipolar dynamics, due to intercell interactions, results in ferroelectric properties of perovskite crystals with temperature-dependent order-disorder phase transitions and a variety of special effects, confirmed experimentally.

[*] Corresponding Author Email: bersuker@cm.utexas.edu.

Among them is the role of the spin in spontaneous polarization, explaining the origin and predicting multiferroic properties and magneto-electric effects. The theory elucidates also the origin of very strong flexoelectricity, permittivity, and electrostriction, enhanced by three-four orders of magnitude due to the PJTE dynamics, and explains the origin of polar nanoregions and relaxor properties of ferroelectrics. Among other features of the JTE and PJTE, important to materials science, we discuss the prediction of a whole class of molecular and solid state compounds with magnetic-dielectric bistability, illustrated by the example of the CuF_3 molecule, and $LiCuO_2$ and $NaCuO_2$ crystals.

Somewhat different looks the application of the PJTE to manipulate the structure and properties of two-dimensional (2D) systems. Here the knowledge of the origin of their buckling as due to the PJTE is used to influence the latter by coordination to other ions or molecular groups, or by redox processes, or by targeted substitutions, all intended to control the buckling, important in applications. Successful realization of this PJTE methodology is demonstrated on a series of 2D and quasi-2D systems.

1. Introduction

Materials science may be defined as *a translation from microscopic foundations of matter to macroscopic performance in applications.* Although there are no established boarders between pure scientific research of materials' properties and their utilization in various applications, it is implied that in materials science the actual and possible applications of the scientific findings are accentuated and paid most attention. In this respect, a rather classical quantum-mechanical effect of specific interactions between the electronic states and nuclear dynamics in different forms of matter, known as the Jahn-Teller effect (JTE) and its extension the pseudo Jahn-Teller effect (PJTE), emerged as important features that strongly influence materials properties with respect to applications. Presently, the JTE and PJTE and the variety of their manifestations in the observable properties of molecular systems and solids are well studied, both theoretically an experimentally (see, e.g., [1-5] and references therein), but it is safe to say that studies of their implications in properties with direct applications in materials science are at the very beginning [6]. This chapter is the first of this kind to summarize the already achieved results of the application of the

JTE and PJTE in materials science. The findings in this trend promise further important applications.

We begin with a brief introduction with definitions of the JTE and PJTE (Section 2) followed by discussion of their main features, especially the induced by these effects specific local dynamics in polyatomic systems, important to the applications (Section 3). Then follow perovskite ferroelectric properties, including order-disorder ferroelectric phase transitions, multiferroicity, magneto-electric effects, and relaxor properties (Section 4). In Section 5 the extremely high sensitivity of materials with JTE and PJTE to external perturbations, leading to giant (enhanced by several orders of magnitude) flexoelectricity, permittivity, and electrostriction, is demonstrated. The next Section 6 is devoted to the dielectric-magnetic bistabilities that occur in molecular and solid state materials due to these effects. Finally, Section 7 describes the practical methodology to influence the PJTE-induced buckling of two-dimensional systems, with a variety of already realized examples.

2. The Physics in the Origin of the Jahn-Teller and Pseudo Jahn-Teller Effects

The vibronic coupling effects, the Jahn-Teller effect (JTE) and the pseudo JTE (PJTE) take place in any polyatomic system if in its given nuclear configuration the electronic state is degenerate (in the case of the JTE), or pseudodegenerate with relatively small energy gaps (in the PJTE case) [1-4]. These effects follow directly from the first-principle (quantum-mechanical) formulation of the problem, in which the coupling between the electronic and nuclear motions (the vibronic coupling), under certain conditions, makes the given (reference) high-symmetry configuration of the system unstable with respect to lower symmetry nuclear displacements [1]. Figure 1 illustrates this situation in the simple case of two electronic states coupled via only one active symmetry-breaking coordinate Q (here and below Q is the defined by group-theory symmetry adapted linear combination of the nuclei's Cartesian displacements [1]) in the case of degeneracy (Figure 1*a*) and pseudodegeneracy (Figure 1*b*). We see that in both cases the ground state in the high-symmetry configuration at Q=0 is unstable (has no minimum of the potential energy) with respect to nuclear displacement of lower symmetry Q, the minima being displaced to nonzero values $Q = \pm Q_0 \neq 0$, where the nuclear configuration is distorted.

Although both kinds of distortions, JTE and PJTE, emerge due to the vibronic coupling, their microscopic driving forces are different. Indeed, quantum-mechanical electronic degeneracy (that leads to the JTE) means that there are two or more different (orthogonal) electronic distributions with the same energy, each of which has lower symmetry than that of the polyatomic system in the reference high-symmetry configuration (with the same symmetry and energy the two electron distributions will not be orthogonal). The configuration distortion takes place because the nuclei follow one of the two lower-symmetry electronic charge distributions, being thus (roughly) an electrostatic effect. In the case of the PJTE the interacting (vibronic coupled) electronic states are relatively close in energy, but not degenerate, and the picture of distortion origin is more complicated; it was shown to be due to gain in covalence bonding by distortion [1, 3]. For clarity and further use in the sections below, we discuss here the origin of the PJTE in more detail.

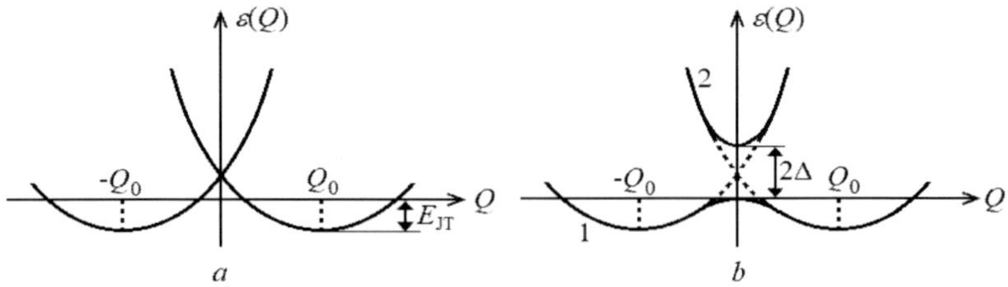

Figure 1. Schematic illustration of the potential energies in the simplest cases of the JTE (*a*) and PJTE (*b*) when two electronic states are coupled via one active displacement mode Q. Distinguished from usual cases of avoiding crossing, the two states in the PJTE have different symmetry, Q is a low-symmetry (distortive) coordinate, and the energy gap 2Δ at Q=0 may be relatively large. E_{JT} is the Jahn-Teller stabilization energy.

The equilibrium geometry of any polyatomic system at T=0 K is defined as corresponding to the point of the minimum of the *adiabatic potential energy surface* (APES), where its first derivatives with respect to nuclear displacements are zero and the second derivatives are positive. This stable configuration does not necessarily coincide with the highest-symmetry or any high-symmetry configuration. We define the latter as the *reference configuration* corresponding to the point of the APES where the first derivatives are zero too, but the second derivative in some direction may be either positive or negative, meaning the high-symmetry reference configuration may be unstable. Denote the energy of the system as a function of normal displacements Q_Γ by $E(Q_\Gamma)$. At the point of minimum of the APES at $Q_\Gamma=0$ the curvature K of $E(Q_\Gamma)$ in the Q_Γ direction,

$$K_\Gamma = (\partial^2 E/\partial Q_\Gamma^2)_0 \quad (1)$$

is positive, $K_\Gamma > 0$. As mentioned above, very often the geometry of the system at this point of equilibrium on the APES does not coincide with the highest possible (or even with any high) symmetry expected from general symmetry considerations. For instance, linear molecules are bent at equilibrium, planar molecules are puckered, octahedral complexes are elongated, or compressed, or tilted, cubic crystals are tetragonally or trigonal polarized (or have several structural phases), etc. The PJTE is the general driving force for all these distortions of high-symmetry configurations in non-degenerate states at T=0 K [1, 3-5].

If at the point of high-symmetry configuration on the APES the system is structurally unstable with respect to some nuclear displacements Q_Γ, than $K_\Gamma < 0$ in this direction. Let us analyze this possibility. The general formula for the energy is $E = \langle\psi_0|H|\psi_0\rangle$, where H is the Hamiltonian and ψ_0 is the wavefunction of the nondegenerate ground state. Substituting E in Eq. (1), we get (omitting the Γ index for simplicity):

$$K=\langle\psi_0|(\partial^2 H/\partial Q^2)_0|\psi_0\rangle +2\langle\psi_0|(\partial H/\partial Q)_0|\psi_0'\rangle \quad (2)$$

where $\psi_0' = (\partial\psi_0/\partial Q)_0$, and all the functions in Eq. (2) are now taken at the point of high-symmetry configuration. Denoting the first term in (2) by K_0, termed the *primary force constant*,

$$K_0 = <\psi_0|(\partial^2 H/\partial Q^2)_0|\psi_0>, \tag{3}$$

and the second one as K_v,

$$K = K_0 + K_v, \tag{4}$$

we notice that $K_v < 0$. Indeed, presenting ψ_0' in terms of the second order perturbation theory, we get directly:

$$K_v = -2\sum_n \frac{|<\psi_0|(\partial H/\partial Q_\alpha)_0|\psi_n>|^2}{E_n - E_0} \tag{5}$$

where ψ_n are the wavefunctions of the excited states and E_n are their energies. This K_v term thus contributes to the instability of the system. Therefore, if $K_0>0$, the K_v contribution is the only source of instability. Note that we define the reference high-symmetry configuration with respect to the displacements Q as the point of the APES at which $(\partial E(Q)/\partial Q)_0 = 0$, meaning zero distortion force in this direction. The matrix elements in Eq. (5) are off-diagonal vibronic coupling constants [1],

$$F_{0n} = <\psi_0|(\partial H/\partial Q)_0|\psi_n>, \tag{6}$$

that mix the ground and excited states under the nuclear displacements Q; therefore K_v is termed *vibronic contribution*. Together with the primary force constant K_0 and the energy gap $2\Delta_{0n} = E_n - E_0$ between the mixing states, F_{0n} are the main parameters of the PJTE.

In a series of papers, beginning in 1980 (see in [1, 3, 4]), it was proved that for any polyatomic system in the high-symmetry configuration

$$K_0 > 0 \tag{7}$$

and hence the vibronic contribution K_v is the only source of instability of any polyatomic system in a high-symmetry configuration in a nondegenerate state. This is the *theorem of instability of high-symmetry configurations*. If $K_0 > 0$ for the high-symmetry configuration (for any polyatomic system), then a negative curvature $K = (K_0 + K_v) < 0$ can be achieved only due to the negative vibronic coupling component K_v, and only if $| K_v | > K_0$. It follows that any distortion of the high-symmetry configuration is due to, and only to the mixing of its ground state with excited electronic states by the distortive nuclear displacements realized via the nonzero vibronic coupling constants (6) in Eq. (5). It means that to rationalize the origin of molecular shapes (or crystal structures) by reasonably starting with some high-symmetry (force-equilibrated) configuration, we notice that there is no way to get this configuration distorted unless we involve its excited electronic states via the vibronic coupling. The latter softens the system with respect to certain nuclear displacements ($K_v < 0$), and if this softening is larger than the original (non-vibronic) stiffness K_0 in this direction, the system becomes unstable with respect to the distortions under consideration, leading to its equilibrium geometry of lower symmetry, or to dissociation.

Thus, starting with a high-symmetry configuration of a polyatomic system and looking for the reason why it may be unstable with respect to distortions (low-symmetry normal nuclear displacements Q), we come to the negative contributions of the excited states K_v to the curvature of the APES in the ground state K_0, given by Eq. (5). It contains a theoretically infinite number of terms including all the corresponding excited states that are mixed with the ground one under the nuclear displacements Q, but practically this number is limited, first by symmetry, and second by energy gaps. Indeed, the matrix elements in the nominators of Eq. (5), the mentioned above vibronic coupling constants $F_{0n} = \langle\psi_0|(\partial H/\partial Q)_0|\psi_n\rangle$, are non-zero only if the product of symmetries of the two wavefunctions contains the symmetry of the symmetrized displacement Q, while the contribution of the higher excited states is strongly diminished by the increasing energy gaps $2\Delta_{0n}$ =

E_n-E_0 in the denominators. As Q is a low-symmetry displacement, the two wavefunction should have different symmetry.

The meaning of the vibronic coupling constant F in Eq. (6) is that it reflects the rate of increase of the overlap of the two wavefunctions (under the Q displacements), which is zero in the high-symmetry configuration. For these reasons only 1, 2, or several low-lying excited states effectively take part in the softening of the high-symmetry configuration of the ground state, and they can be revealed in a two-level (or n-level) problem. The symmetry of the most active excited state defines also the distortion coordinate Q, discriminated among several possible in the high-symmetry configuration. It is important that because of the off-diagonal character of the constant F, the increase of wavefunctions' overlap by distortion includes mainly near-neighbor atomic functions' overlap due to the changed in interatomic distances. In other words, the *driving force of the PJTE is the gain of energy due to the improved covalence bonding between the atoms in the distorted configuration* [1, 3].

3. Exclusive Features of the JTE and PJTE Important to Materials Science: Dynamics of Local Distortions

The briefly outlined above main physical outcome from the JT and PJT effects is the instability of the high-symmetry configuration of polyatomic systems, but the problem of interest is the influence of this instability on observable properties. The consequences of the JT and PJT effects, their influence on properties, described in an "ocean of publications", are huge (see, e.g., [1-6]). In addition to the expected distorted configurations induced by these effects, there are some special features, which were shown recently to strongly influence the properties (and create new properties) of materials, leading to important applications. The main of these features can be termed as the *dynamics of local distortions.* We illustrate this feature with two examples for the two effects, JTE and PJTE, respectively.

Consider one of the simplest and most widespread case of the JTE, when the electronic term of the system (a molecule, cluster, or local center in solids) in a given (high-symmetry) nuclear configuration is double degenerate, an *E* term. According to the theory [1, 2, 4], the vibronic coupling makes this configuration unstable with respect to two-fold degenerate *e* type normal nuclear displacements Q_ϑ and Q_ε, so the JTE problem is denoted as $E\otimes e$. The solution of the rather simple secular equation for this problem yields the dependence of the energy of the system on the nuclear coordinates, the adiabatic potential energy surface (APES), which in polar coordinates, $Q_\vartheta = \rho\cos\phi$, and $Q_\varepsilon = \rho\sin\phi$ (up to quadratic terms) looks as follows [1]:

$$\varepsilon_{\pm}^{v}(\rho,\phi) = \frac{1}{2}K_E\rho^2 \pm \rho[F_E^2 + G_E^2\rho^2 + 2F_E G_E\rho\cos 3\phi]^{\frac{1}{2}} \qquad (8)$$

Here F_E and G_E are the linear and quadratic vibronic coupling constants, respectively, and K_E is the introduced in Eq. (3) primary force constant for the *e* type nuclear displacements. The surface (8) is illustrated in Figure 2. We see that the APES in this case has three equivalent minima divided by potential barriers. In each of these minima, the system is distorted, as shown schematically in Figure 3 for an octahedral and tetrahedral system. All the details of this APES are well described analytically [1-4]. For the purpose of the applications below in this paper, we take into consideration that without external perturbations the ground state in such systems is an average between three equivalent, but differently oriented distortions, the averaging being achieved by tunneling or hoping, dependent on the height of the barriers between the minima of the APES. These *dynamic distortions* localized around the JTE center of the systems is the most important feature of the JTE with direct applications in materials science. Note that in a time scale these dynamics are much slower (by orders of magnitude) than typical vibrations.

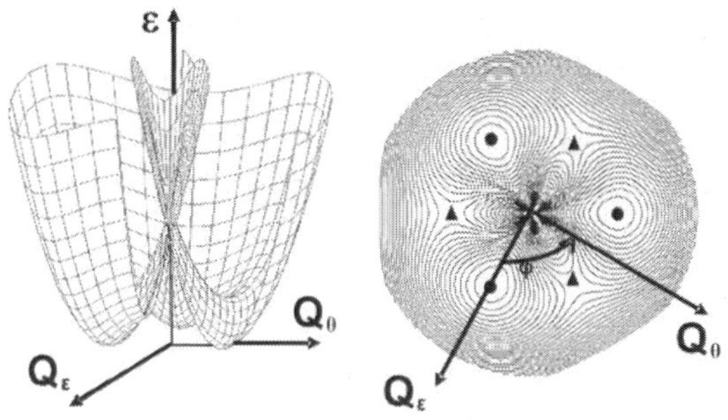

Figure 2. The APES for the JTE problem $E \otimes e$ with both the linear and quadratic terms of the vibronic interaction included: general view (left, the "*tricorn*"), and the equipotential sections of the lower branch (right); three minima and three saddle points are indicated by black circles and triangles, respectively. (Reprinted with permission from Ref. [1]. Copyright 2006, I. B. Bersuker).

As mentioned above, in the case of the PJTE the picture of distortions and their dynamics is more complicated. It can be best illustrated for a specific system, widely used below in applications. As the PJTE in crystals is of local origin, consider the $[BO_6]^{8-}$ cluster in a perovskite crystal ABO_3, e.g., $BaTiO_3$.

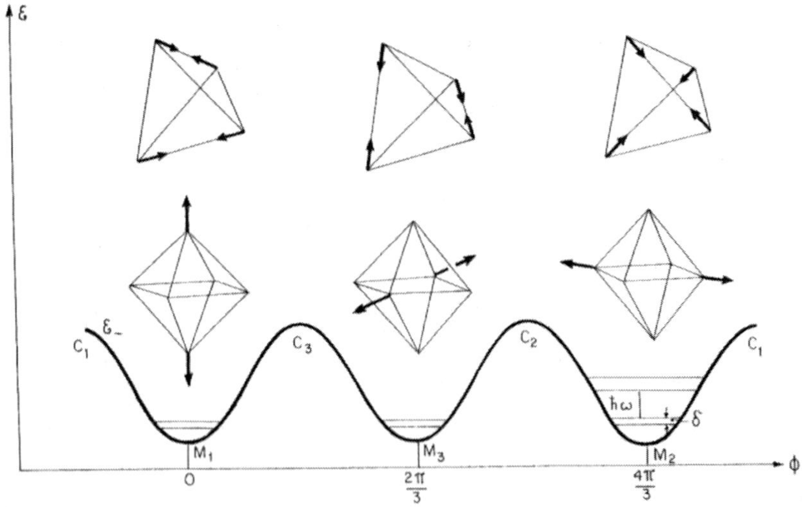

Figure 3. The section of the lowest branch ε_- of the APES of the quadratic $E \otimes e$ problem along the line of steepest slope with illustration of the distortions of octahedral and tetrahedral systems at the points of minima, and the tunnelling splitting in each of them.

The octahedral fragment of the unit cell centered at the transition metal atom B is shown in Figure 4. It is surrounded by six oxygen atoms (blue circles in Figure 4; cation atoms A are not shown). In the molecular orbital (MO) presentation of this octahedral fragment [7-9], the highest occupied MO (HOMO) are formed by the atomic $2p_\pi$ orbitals of the six oxygen atoms. They belong to the representations T_{1u}, T_{2u}, and T_{1g} of the octahedral symmetry group and are respectively denoted in Figure 5. The lowest unoccupied MO (LUMO) are $t_{2g}(3d)$ orbitals of the transition metal: $|yz(B)\rangle$, $|xz(B)\rangle$, and $|xy(B)\rangle$. In the regular cubic configuration, these HOMO-LUMO orbitals are orthogonal and do not participate directly in the Ti—O bonding (the bonding is realized mainly via the inner σ orbitals and ionic interactions). Obviously, any off-center displacement of the atom B lowers the cubic symmetry and makes their overlap nonzero. Figure 6 illustrates this effect for the orbital $|yz(Ti)\rangle$.

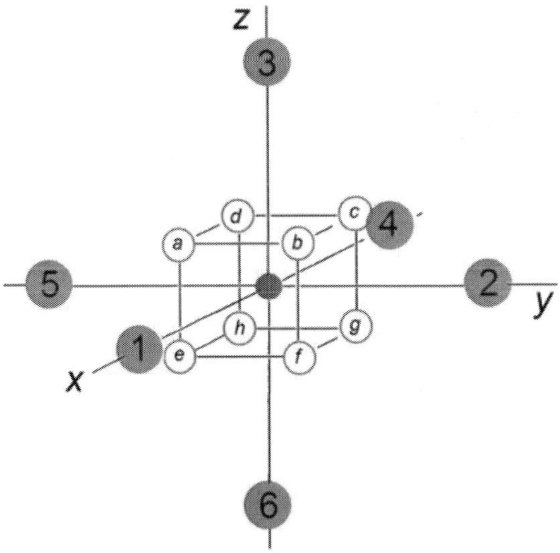

Figure 4. (Color online). The octahedral fragment of the perovskite crystal structure ABO_3 with the transition metal atom B at the center (red) and six oxygen atoms at the apexes of the octahedron (numbered, blue). The letters a, b, c ...denote the induced by the PJTE eight equivalent off-center positions of the atom B in the eight wells of the APES. [Reprinted with permission from Ref. [8]. Copyright 2015 Elsevier Publishing].

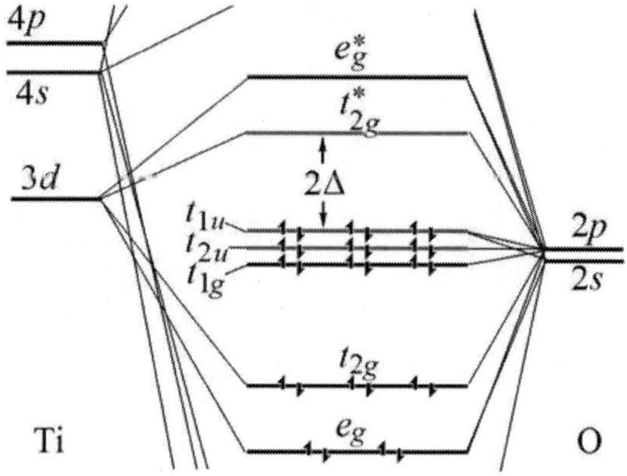

Figure 5. The energy-level correlation diagram (not to scale) for the octahedral cluster $[TiO_6]$. HOMO t_{1u} and t_{2u} (mostly oxygen) and LUMO t_{2g} (mostly titanium) are one-electron energy levels of the orbitals that are mixed in the PJTE. The shown electron population corresponds to the particular case of $[TiO_6]^{8-}$ [1, 8, 9]. [Reprinted with permission from Ref. [1]. Copyright 2006, I. B. Bersuker].

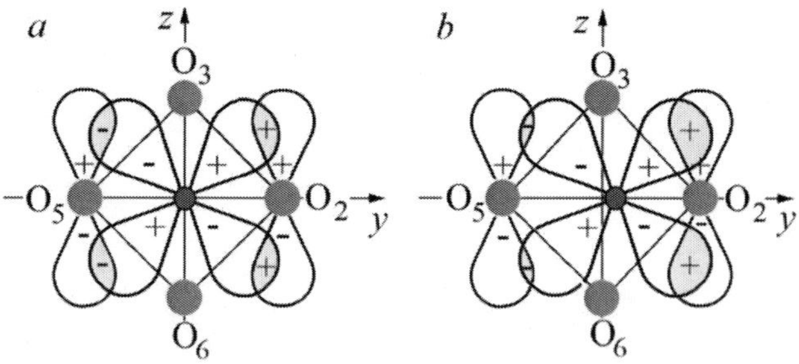

Figure 6. (Color online). Added covalence due to low-symmetry distortions induced by the PJTE. Shown in yellow is the overlap of the 3d orbital $|T_{2g}yz(Ti)\rangle$ with the symmetry-adapted π orbital $|T_{1u}z\rangle$ of the surrounding oxygen atoms. (a) When Ti atom is at the center, the positive contribution to the overlap cancels the negative one, so the net overlap is zero. (b) When the atom of Ti is off-center shifted in the y-direction, the net overlap gains a non-zero value. [Reprinted with permission from Ref. [8]. Copyright 2015 Elsevier Publishing].

The vibronic PJT coupling between the ground electronic state (the HOMO) with the excited state (LUMO) by the nuclear T_{1u} displacements Q_α, α = x, y, z, of the cluster $[BO_6]^{8-}$, eased by the high symmetry and closed-shell electronic structure (which reduces the problem to a matrix

9 × 9), yields the APES, $U(Q_x,Q_y,Q_z,)$, in the analytic form, obtained already in the first paper of the PJTE theory [7] (see also [1, 8, 9]) as follows:

$$U(Q)=\tfrac{1}{2}K_0Q^2-2\left[\sqrt{\Delta^2+2F^2\left(Q^2-Q_x^2\right)}+\right.$$
$$\left.+\sqrt{\Delta^2+2F^2\left(Q^2-Q_y^2\right)}+\sqrt{\Delta^2+2F^2\left(Q^2-Q_z^2\right)}\right]. \tag{9}$$

Here $Q^2 = Q_x^2 + Q_y^2 + Q_z^2$, 2Δ is the (band adapted [8]) energy gap between the PJT active ground and excited electronic states, $\frac{1}{2}K_0Q^2$ is the added elastic energy with respect to these displacements, K_0 is the respective primary (non-vibronic) force constant, characterizing the crystal stiffness without the vibronic coupling (presented as a crystalline multimode sum with the expansion of Q over the corresponding crystal vibrations [8]), and F is the vibronic coupling constant,

$$F=\left\langle 2p_y(\mathrm{O})\left|\left(\frac{\partial H}{\partial X}\right)_0\right|3d_{xy}(\mathrm{Ti})\right\rangle \tag{10}$$

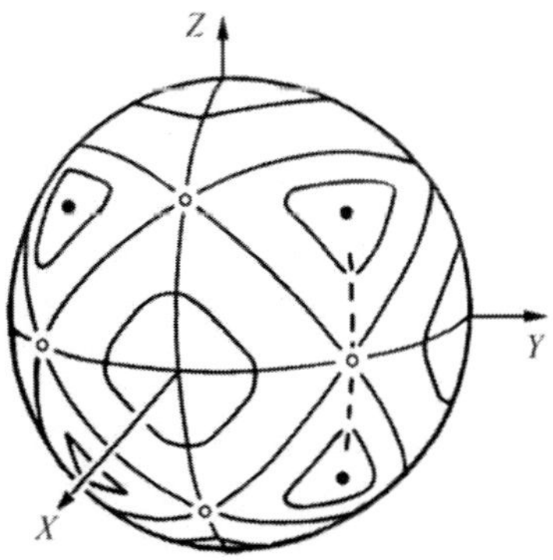

Figure 7. Contour map of the ground-state APES $U(Qx,Qy,Qz)$ for the octahedral cluster $[BO_6]^{8-}$ due to the PJT coupling of its oxygen HOMOs to the central-atom's LUMOs. Equipotential level curves are projected on the surface of a sphere. Solid dots represent trigonal (rhombohedral) minima, open dots are orthorhombic saddle points. Broken curve is the path of steepest descent from one of the saddle points to the closest wells.

The APES (9) has interesting features: under the condition

$$\Delta < \frac{8F^2}{K_0} \tag{11}$$

$U(Q)$ has a maximum (instability) with the B ion in the center, eight equivalent minima situated along the eight trigonal directions toward the cubic vertices (at the points a, b, c, …in Figure 4), twelve equivalent saddle points along the six second-order symmetry axes C_2 (in two directions from the center), each of them at the top of the energy barrier between two near-neighbor equivalent minima, and higher in energy six equivalent saddle-points along the three tetragonal C_4 axes (in two directions), each of them forming a higher barrier between the groups of four near-neighbor equivalent minima. Figure 7 illustrates some of the main features of the surface.

Table 1. Numerical estimates of the parameter values of the local APES (9) induced by the PJTE for the Ti center in $BaTiO_3$ [8]

Main parameters	**Assumed: Δ = 1.4 eV, $\hbar\omega_E$ = 193 cm^{-1}**
δ	6.6
K_0	55 eV/Å^2
Δ	1.4 eV
F	3.42 eV/Å
F^2/K_0	0.21 eV
F/K_0	0.06 Å
$\hbar\omega_0$	684 cm^{-1}
$\hbar\omega_E$	193 cm^{-1}
$\lvert Q_x\rvert = \lvert Q_y\rvert = \lvert Q_z\rvert$	0.14 Å
$Q_x = Q_y = q_0, Q_z = 0$	q_0 = 0.16 Å
E_{JT}[000]	0
E_{JT}[100]	880 cm^{-1}
E_{JT}[110]	1130 cm^{-1}
E_{JT}[111]	**1250 cm^{-1}**

The APES (9) leads to a series of important properties with direct applications in materials science, discussed below in this paper. Almost all of them follow from the basic feature of the multiminimum potential energy as a function of nuclear displacements. *The multiminimum character of the APES and consequent dynamics of the distortions is an unalienable property of polyatomic systems with the JTE and PJTE* in the absence of external perturbations.

A more accurate quantitative treatment of the problem was achieved employing the Green's function theory for the local PJTE vibronic coupling [8], which takes into account the contribution of the whole crystal (via its electronic and vibrational bands) to the local distortion. It allowed us to estimate the main parameters for the most studied and used perovskite ferroelectric crystal, $BaTiO_3$, involving only one empirical parameter, the band-adapted energy gap between the PJT active ground and excited electronic states, $2\Delta = 2.8$ eV, and the DFT calculated vibrational frequency in the minima of the APES $\hbar\omega_E = 193$ cm^{-1} [8]. Table 1 shows the other parameters values of the surface (9) plus the coordinates of the trigonal minima and the saddle point between them, as well as their energies.

4. PJTE IN THE ORIGIN OF FERROELECTRICITY

4.1. Local Dipolar Distortions and Order-Disorder Phase Transitions

The described above APES (Eq. (9)), which occurs under certain conditions at the B centers of ABO_3 perovskite crystals due to the PJTE, emerges as a result of quantum-mechanical calculations (at T = 0) based on first principles. The APES remains the same at higher temperatures, but the observable properties depend on the nuclear motions with this APES, which are strongly temperature dependent. As a function of nuclear coordinates with its maximum in the center, eight trigonal minima, 12 orthorhombic saddle-points, and six higher-in-energy tetragonal saddle-points, the APES (9) triggers a series of important properties relevant to materials science (similar considerations apply to cases when the A center of the ABO_3

perovskite is active). For a review of the vibronic PJT theory of ferroelectricity, see Ref. [9].

To begin with, the APES in Eq. (9) reveals the origin of all the four temperature dependent phases of such ferroelectrics and the order-disorder phase transitions between them, described already in the first paper on this theory [7]. In the free cluster $[BO_6]^{8-}$ the eight possible off-center equilibrium positions of the ion B with local dipole moments are equivalent, but in the crystal they may become nonequivalent due to intercell interactions and the mean field E of the environment to each center induced by the off-center positions of the B ion in the near-neighbor cells. *Both the local off-center displacement and the mean field E of the environment are interdependent in a self-consistent way* [10]. At T = 0 K the lowest free energy is achieved when all the B ions occupy the trigonal minima in the same direction, say [111], and the crystal is in its low-fully ordered rhombohedral phase.

With temperature, Boltzmann population of excited energy levels grows. When the temperature excitation in the trigonal wells approaches the height of orthorhombic potential barriers, the system averages over two nearest-neighbor trigonal wells along, say, [111] and [11-1], by hoping over, or tunneling through, the orthorhombic barrier along [110]. The crystal becomes disordered along one of its symmetry axes, the z-axis in this example, remaining ordered along the other two. In average, both the mean field E and the polarization are oriented along [110]. Macroscopically, this type of ordering is seen as the orthorhombic phase. Still, lowered by the mean field, this pair of potential wells is separated from the other equivalent ones by tetragonal potential barriers that are higher in energy. By a farther increase in temperature, higher-energy states become temperature populated, their energy reaching the higher potential barrier with the tetragonal saddle point with, say, [100] on its top. Figure 8 illustrates the APES minima positions in such nonzero mean field $E \neq 0$. The crystal structure transforms into the tetragonal phase with two directions of disorder (four near-neighbor minima of the APES involved). The average polarization and the induced mean field E change to the one along [100]. This is the tetragonal phase.

Finally, when the temperature is high enough, the maximum of the adiabatic potential at $Q_x = Q_y = Q_z = 0$ is overcome (or sidestepped [10]). The local dipole moments become (macroscopically) averaged over all eight trigonal wells, all three directions, and the crystal reaches the completely disordered paraelectric phase. Predicted in 1966 [7], this picture of the induced by the PJTE one ordered, two partially disordered, and one fully disordered phases in ferroelectric perovskites, and the order-disorder phase transition between them, was first observed experimentally in 1968 in $BaTiO_3$ and $KNbO_3$ by means of X-ray diffuse scattering [11] (see also [12]). Many consequent experimental observations and further theoretical developments (see the review [9]) confirm this conclusion. A rigorous treatment of the phase transitions in view of the dynamic PJTE local polar distortions, carried out in the mean field approximation, yields reasonable temperatures of the phase transition [10]. The emerged vibronic (PJT) theory of perovskite ferroelectricity is reviewed in Ref. [9].

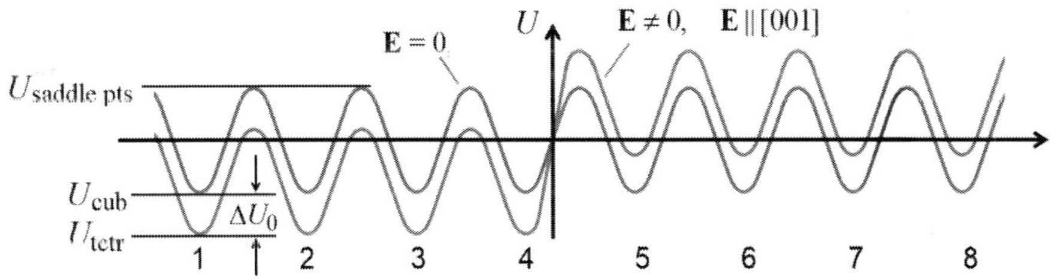

Figure 8. (Color online). Schematic cross-section of the APES $U(Q_x,Q_y,Q_z,)$ along the path of steepest descent from the orthorhombic saddle-points to the trigonal minima. At $T > T_C$, in the cubic phase the mean field equals zero, E = 0, and all the eight trigonal minima have the same depth (rose curve, upper for 1,2,3,4, and lower for 5,6,7,8). At $T < T_C$ (green curve, lower for 1,2,3,4, and upper for 5,6,7,8) the mean field E is oriented along one of the tetragonal axes (here along [001]) lowering the energy of four minima 1, 2, 3 and 4 (*a,b,c,d* in Figure 1) and elevating the energy of the other four minima. ΔU_0 is the energy gain by ordering.

A question emerges, can the *instant* dipolar displacements of the B ion be observed directly? The answer to this question is related to yet another feature of the ferroelectrics under consideration that has nothing to do with displacive theories and get full explanation in the vibronic theory, the *relativity to the means of observation.* Similar to many other JTE and PJTE systems with multiminimum APES, there may be differences in experimental observations of the same phenomena with different methods [5]. This effect can be attributed to the significant differences in the "time-scale" and "space-scale" of different methods of observation. With regard to the time-scale, different experimental methods of observation have different "time of measurement" τ', which may be comparable with the time $\tau \approx \hbar/\Gamma$ of the reorientational hoping of the metal ion between the minima of the APES (see in [9]; Γ is the tunneling or hopping frequency). The conclusions from the experimental data may be different for experiments with $\tau' > \tau$ and $\tau' < \tau$ because of the full or partial averages over the reorientation positions (APES minima) observed in the former case [5, 9].

In cubic perovskites Γ ranges from 0.01 cm^{-1} to 50 cm^{-1}. Respectively, the characteristic time is from $\tau \approx 10^{-9}$ s = 1 ns to $\tau \approx 10^{-13}$ s = 0.1 ps. For the tunneling via the lowest barrier in $BaTiO_3$ estimates obtained by EPR experiments with ion substitution yield $\tau\sim 10^{-9}\text{-}10^{-10}$ s [40, 41]. For instance, the origin of apparent differences in the interpretation of the experiments on NMR [13] and XAFS [14] at the tetragonal-to-cubic phase transition in $BaTiO_3$ were explained as due to the difference in the time-scales of the two methods of measurement [14]: in NMR $\tau' \sim 10^{-8}$ s, which is larger than $\tau \sim 10^{-9}\text{-}10^{-10}$ s, and thus shows an averaged picture [15], whereas in XAFS $\tau' \sim 10^{-15}$ s, which is smaller than τ and reveals the distorted configuration.

Similarly, for the "space-scale" factor, the conclusions from the experiments may depend on the relation between, e.g., the wavelengths in optical measurements as compared with the unit cell dimensions l, where the dipole averaging occurs; for barium titanate it is of the order of $l \approx 4$ Å = 0.4 nm. All optical methods, including IR absorption, hyper-Raman, or Brillouin scattering of light, are long-range methods. Their experimental length, l_{exp}, is of the order of wavelength of the incident or scattered light, $l_{exp} \sim 10$ nm ÷ 10 μm. This is by several orders in magnitude longer than the crystal characteristic size, where the local disorder takes place, $l \approx 0.4$ nm. Therefore, the long-range methods provide the information averaged over large numbers of unit cells.

In view of this scale factor in the relativity to the means of observation, the interpretation of the experimental data on the ferroelectric phase transitions in perovskite crystals acquires additional lightening. Of particular interest is the observations of the low-frequency (soft) mode in neutron scattering, IR absorption, hyper-Raman scattering, etc., at the tetragonal to cubic transition in barium titanate, deemed to show its partial displacive nature [16]. These methods reflect the picture of the crystal with the averaged in time and space local unit-cell disorder. In these methods the inequalities $\tau' >> \tau$ and $l_{exp} >> l$ reach several orders of magnitude. Alternative experiments with diffuse X-ray scattering [11] and EXAFS [14] with $\tau < 10^{-15}$ s and $l_{exp} < 0.01$ nm (hence $\tau' << \tau$ and $l_{exp} << l$), clearly reveal the order-disorder nature of the phase transitions in perovskite ferroelectrics. Significantly, the local B center disorder explains also the formation of "polar nanoregions" (see below).

4.2. Spin Implication, Multiferroicity, and Magneto-Electric Effects

Coexistence of ferroelectric and magnetic properties (multiferroicity) implies that the ferroelectric crystal, which is a dielectric, has also a nonzero magnetic moment, meaning unpaired electrons. In the ferroelectric $BaTiO_3$ with perovskite ABO_3 structure the d^0 configuration of the Ti^{4+} ion has no unpaired electrons, and attempts to obtain ferroelectricity in perovskites with d^n, n > 0, transition metals B ions were unsuccessful for a long time. This prompted some authors to term the situation as a ferroelectric "*d^0 mystery*" accompanied by a conclusion that nonzero spin states are detriment to ferroelectric polarization. However, more recently quite a number of ferroelectrics-multiferroics, mostly perovskites with configurations d^3-d^7, were obtained and studied (see, e.g., in [17]). The origin of these special properties of perovskite ferroelectrics with d^n configurations does not follow directly from displacive theories, and had no general explanation for a long time. The vibronic (PJTE) theory of ferroelectricity elucidates the role of spin states in the local polar instability, explains directly the origin of perovskite multiferroics with proper ferroelectricity, and formulates the necessary conditions that ABO_3 perovskites with a magnetic d^n configuration of the B ion may be ferroelectric [18]. Moreover, because of the spin implication, the multiferroics conditions that emerge from the PJTE for d^n ions with n = 3, 4, 5, 6, and 7, are directly influenced by the well-known transition metal high-spin/low-spin crossover, resulting in the coexistence of three phenomena: ferroelectricity (FE), magnetism (M), and spin crossover (SCO). This, in turn, leads to a quite novel phenomenon: *magnetic-ferroelectric (multiferroics) crossover* (MFCO), creating a rich variety of possible magnetoelectric and related effects [18]. *The vibronic (PJTE) theory, exclusively, reveals the role of spin in spontaneous polarization of crystals.*

The typical molecular-orbital (MO) energy scheme for an octahedral cluster BO_6^{8-} in the approximation of linear combinations of atomic orbitals is discussed above (Section 3) and shown in Figure 5 for the MO electron population of the d^0 configuration, e.g., when B≡Ti. The highest occupied MO (HOMO) in this case is t_{1u} which is a threefold degenerate odd-parity (ungerade) linear combination of mostly oxygen p_π orbitals, while the lowest unoccupied MO (LUMO) is t_{2g}, mostly atomic three d_π orbitals of the transition metal ion B, and the next excited MO is the double degenerate one e_g (the non-bonding oxygen b_{1g} MO is not shown as irrelevant). For the d^0 case the HOMO configuration is thus $(t_{1u})^6 = (t_{1u}\downarrow)^3(t_{1u}\uparrow)^3$, where the arrows up and down indicate the two spin states; the energy term of this configuration is $^1A_{1g}$. The excited state with opposite parity is formed by the one-electron, $(t_{1u}\uparrow) \rightarrow (t_{2g}\uparrow)$ or $(t_{1u}\uparrow) \rightarrow (e_g\uparrow)$, excitation, resulting in the lowest excited ungerade term $^1T_{1u}$ at the energy gap 2Δ. In this case, the PJTE at the B center, under the condition of instability $\Delta < 8F^2/K_0$ (Eq. (11)), produces a polar displacements of the B atom along [111] type directions, which triggers the ferroelectric polarization (Section 4).

For the d^1 configuration of the B ion instead of the d^0, the HOMO becomes $(t_{1u}\downarrow)^3(t_{1u}\uparrow)^3(t_{2g}\uparrow)^1$ with the term $^2T_{2g}$, and LUMO (taken into account Hund's rule) is $(t_{1u}\downarrow)^2(t_{1u}\uparrow)^3(t_{2g}\uparrow)^2$ with the lowest excited ungerade term $^4T_{1u}$. Hence the two closest terms of different parity, *possess different spin multiplicity*, and hence *they do not mix by the vibronic coupling*; the latter does not contain spin operators [1, 2, 4]. In principle, there may be higher in energy electronic configurations of opposite parity with the same spin as the ground state one, but they are at much larger energy gaps Δ, and therefore less appropriate to satisfy the condition of instability (11) (numerical estimates in Section 4 show that the condition (11) is very restricting). A similar picture emerges for d^2 configurations for which the two lowest terms of opposite parity are $^2T_{2g}$ and $^5T_{1u}$ (see Table 2).

The situation changes for d^3. Indeed, in this case the HOMO is $(t_{1u}\downarrow)^3(t_{1u}\uparrow)^3(t_{2g}\uparrow)^3$ with the ground state term $^4A_{1g}$, and in the low-spin conditions of the strong ligand fields (large t_{2g}-e_g separation in Figure 2, see, e.g., [19]), the LUMO is $(t_{1u}\downarrow)^2(t_{1u}\uparrow)^3(t_{2g}\uparrow)^3(t_{2g}\downarrow)^1$ with the lowest ungerade term $^4T_{1u}$. Therefore for d^3 configurations (e.g., Mn^{4+}) in sufficiently strong ligand fields the situation becomes again favorable for the PJTE and polar distortions, but in this case, distinguished from the d^0 case, the system possess also a magnetic moment created by three unpaired electrons. However, if the ligand field is weak and the separation t_{2g}-e_g is small, the high-spin arrangement of the excited electronic configuration takes place, and the excitation electron occupies the $e_g\uparrow$ orbital instead of $t_{2g}\downarrow$; the LUMO configuration under Hund's rule becomes $(t_{1u}\downarrow)^2(t_{1u}\uparrow)^3(t_{2g}\uparrow)^3(e_g\uparrow)^1$ with the lowest ungerade state $^6T_{1u}$. Here again there is no PJTE on dipolar distortions and no ferroelectric instability. This is one of the examples, which shows explicitly how the spin states interfere directly into the possible local polar displacement and ferroelectricity. In this way all the d^n configurations with n = 0, 1, 2, ...10, were explored, the results being shown in Table 2.

It follows from the table 2 that according to the PJTE, in perovskite ABO_3 crystals only B ions with configurations d^0, d^3-low-spin, d^4-low-spin, d^5-low-spin and high-spin, d^6-high-spin and intermediate-spin, d^7-high-spin, d^8, and d^9 can, in principle, produce multiferroics, provided the criterion of instability (11) is satisfied.

Table 2. Necessary conditions that ABO_3 perovskites with the electronic d^n configuration of the B ion possess both ferroelectric and magnetic properties simultaneously [18]

d^n	*Example*	*HOMO EC and GS term*	*LUMO EC and LUES term*	*FE*	*MM*	*MF*
d^0	Ti^{4+}	$(t_{1u})^6$, $^1A_{1g}$	$(t_{1u})^5(t_{2g}\uparrow)^1$, $^1T_{1u}$	yes	no	no
d^1	Ti^{3+}, V^{4+}	$(t_{1u})^6(t_{2g}\uparrow)^1$, $^2T_{2g}$	$(t_{1u})^5(t_{2g}\uparrow)^2$, $^4T_{1u}$	no	yes	no
d^2	V^{3+}, Cr^{4+}	$(t_{1u})^6(t_{2g}\uparrow)^2$, $^3T_{2g}$	$(t_{1u})^5(t_{2g}\uparrow)^3$, $^5T_{1u}$	no	yes	no
d^3, LS	Cr^{3+}, Mn^{4+}	$(t_{1u})^6(t_{2g}\uparrow)^3$, $^4A_{2g}$	$(t_{1u})^5(t_{2g}\uparrow)^3(t_{2g}\downarrow)^1$, $^4T_{1u}$	yes	yes	yes
d^3, HS		$(t_{1u})^6(t_{2g}\uparrow)^3$, $^4A_{2g}$	$(t_{1u})^5(t_{2g}\uparrow)^3(e_g\uparrow)^1$, $^6T_{1u}$	no	yes	no
d^4, LS	Mn^{3+}, Fe^{4+}	$(t_{1u})^6(t_{2g}\uparrow)^3(t_{2g}\downarrow)^1$, $^3T_{2g}$	$(t_{1u})^5(t_{2g}\uparrow)^3(t_{2g}\downarrow)^2$, $^3T_{1u}$	yes	yes	yes
d^4, HS		$(t_{1u})^6(t_{2g}\uparrow)^3(e_g\uparrow)^1$, $^5T_{2g}$	$(t_{1u})^5(t_{2g}\uparrow)^3(e_g\uparrow)^2$, $^7T_{1u}$	no	yes	no
d^5, LS	Mn^{2+}, Fe^{3+}	$(t_{1u})^6(t_{2g}\uparrow)^3(t_{2g}\downarrow)^2$, $^2T_{2g}$	$(t_{1u})^5(t_{2g})^6$, $^2T_{1u}$	yes	yes	yes
d^5, HS		$(t_{1u})^6(t_{2g}\uparrow)^3(e_g\uparrow)^2$, $^6A_{1g}$	$(t_{1u})^5(t_{2g})^4(e_g\uparrow)^2$, $^6T_{1u}$	yes	yes	yes
d^6, LS	Fe^{2+}, Co^{3+}	$(t_{1u})^6(t_{2g})^6$, $^1A_{1g}$	$(t_{1u})^5(t_{2g})^6(e_g\uparrow)^1$, $^3T_{1u}$	no	no	no
d^6, IS_1		$(t_{1u})^6(t_{2g})^5(e_g\uparrow)^1$, $^3T_{1g}$	$(t_{1u})^5(t_{2g})^6(e_g\uparrow)^1$, $^3T_{1u}$	yes	yes	yes
d^6, IS_2		$(t_{1u})^6(t_{2g})^5(e_g\uparrow)^1$, $^3T_{1g}$	$(t_{1u})^5(t_{2g})^5(e_g\uparrow)^2$, $^5T_{1u}$	no	yes	no
d^6, HS		$(t_{1u})^6(t_{2g})^4(e_g\uparrow)^2$, $^5T_{2g}$	$(t_{1u})^5(t_{2g})^5(e_g\uparrow)^2$, $^5T_{1u}$	yes	yes	yes
d^7, LS	Co^{2+}, Ni^{3+}	$(t_{1u})^6(t_{2g})^6(e_g\uparrow)^1$, ^2E_g	$(t_{1u})^5(t_{2g})^6(e_g\uparrow)^2$, $^4T_{1u}$	no	yes	no

d^n	*Exa-mple*	*HOMO EC and GS term*	*LUMO EC and LUES term*	*FE*	*MM*	*MF*
d^7, HS		$(t_{1u})^6(t_{2g})^5(e_g\uparrow)^2$, ${}^4T_{2g}$	$(t_{1u})^5(t_{2g})^6(e_g\uparrow)^2$, ${}^4T_{1u}$	yes	yes	yes
d^8	Ni^{2+}, Cu^{3+}	$(t_{1u})^6(t_{2g})^6(e_g\uparrow)^2$, ${}^3A_{1g}$	$(t_{1u})^5(t_{2g})^6(e_g)^3$, ${}^3T_{1u}$	yes	yes	yes
d^9	Cu^{2+}	$(t_{1u})^6(t_{2g})^6(e_g)^3$, ${}^2E_g$	$(t_{1u})^5(t_{2g})^6(e_g)^4$, ${}^2T_{1u}$	yes	yes	yes
d^{10}	Zn^{2+}	$(t_{1u})^6(t_{2g})^6(e_g)^4$, ${}^1A_g$	$(t_{1u})^5(t_{2g})^6(e_g)^4(ns\uparrow)^1$, ${}^3T_{1u}$	no	no	no

EC = electronic configuration, GS = ground state, LUES = lowest ungerade excited state, FE = ferroelectric, MM = magnetic, MF = multiferroic, LS = low-spin, HS = high-spin; IS = intermediate spin. $(t_{1u})^6 = (t_{1u}\downarrow)^3(t_{1u}\uparrow)^3$; $(t_{1u})^5 = (t_{1u}\downarrow)^2(t_{1u}\uparrow)^3$; $(t_{2g})^6 = (t_{2g}\uparrow)^3(t_{2g}\downarrow)^3$; $(e_g)^4 = (e_g\uparrow)^2(e_g\downarrow)^2$.

If the (in general, hardly significant) contribution of higher excited states can be ignored, transition metal ions B with configurations d^1, d^2, d^3-high-spin d^4-high-spin, d^6-low-spin, d^7-low-spin, and d^{10} are not expected to produce multiferroics under this mechanism of proper ferroelectricity. Experimentally observed perovskite multiferroics with such B ions, for example, $Mn^{4+}(d^3)$, $Cr^{3+}(d^3)$, $Mn^{3+}(d^4)$, $Fe^{3+}(d^5)$, $Fe^{2+}(d^6)$, $Co^{2+}(d^7)$ [17], etc., fit well with the conclusions of Table 3; there are no multiferroics with d^0, d^1, d^2, and d^{10} configurations.

Of special importance is also the fact that some d^n ions with n = 3, 4, 5, 6, and 7, dependent on the ligands of the octahedral environment, may produce two types of magnetic centers, high-spin (HS) and low-spin (LS), and in d^6 case there maybe also intermediate spin (IS) states (d^3 has two spin configurations in the one-electron excitation). According to our analysis, only d^5 systems follow the necessary condition of potential multiferroics in both spin states, but the PJTE condition of instability (11) and the magnetic moments are different in these two cases. For d^3, d^4, d^6, and d^7 ions only one of the two spin states may serve as a candidate of potential multiferroics.

On the other hand, in many cases the two spin states are close in energy, producing a well-known phenomenon of transition metal *spin crossover* (SCO), in which case the system can be relatively easily transferred from one spin state to another by external perturbations like heat, light, and magnetic fields (for reviews see [20]). Since, as shown above, the change of the spin state changes also the possibility of ferroelectric polarization, *the SCO is simultaneously a magnetic-ferroelectric (multiferroic) crossover* (MFCO). This coexisting magnetic, ferroelectric, and spin-crossover phenomena opens a variety of new possibilities to manipulate the properties of the system, with exciting novel functionalities to electronics and spintronics. Here are several examples: 1) For d^3 and d^4 (Cr^{3+}, Mn^{4+}, Mn^{3+}, Fe^{4+}, etc.) ferroelectrics in the LS state in conditions of MFCO, magnetic fields facilitate the LS→HS transition that destroys the ferroelectricity (and multiferroicity), while an electric field in the HS non-ferroelectric state may transfer the system to the ferroelectric (multiferroic) LS state, thus realizing *electric demagnetization*; 2) For d^5 ferroelectrics in conditions of MFCO, if the ferroelectricity is (most probable) different in the two spin states, an electric field may change the spin state (*electric magnetization or demagnetization*); 3) For d^6 and d^7 (Fe^{2+}, Co^{3+}, Co^{2+}, Ni^{3+}, etc.) in the non-ferroelectric LS state under conditions of MFCO, magnetic fields facilitates the LS→HS transition that induces ferroelectricity and hence multiferroicity in a *strong magnetoelectric effect* (the d^6 LS state is nonmagnetic); in the non-ferroelectric LS state in MFCO conditions, an electric field may transfer the system to the multiferroic state (*electric magnetization*); 4) The SCO phenomenon is well known to be influenced also by stress, heat, light, and cooperative effects in crystals [20], hence these perturbations can be used to manipulate the MFCO and all the consequent properties including those mentioned above. The dependence of the MFCO on pressure adds a ferroelastic order to the magnetic and ferroelectric ones; 5) There is already a long history of attempts to use transition metal SCO systems as units of magnetic bistability; the difficulty is in the fast relaxation (short lifetime) of the higher in energy spin state [21]. By choosing a system in the MFCO condition, one can increase the lifetime of the excited dipolar (multiferroic) state by applying an external electric field; 6) An important feature of the revealed MFCO is that it is of local origin and hence it does not necessarily require strong cooperative interactions, meaning that, in principle, it may take place as a magnetic-dipolar effect in separate molecular systems, clusters, thin films, etc., provided the condition of instability (11) is obeyed.

In addition to the mentioned above examples that confirm the (outlined by the PJTE theory) necessary conditions of multiferroicity [17], several papers [22-24] reported observing also the predicted [18] magnetic-ferroelectric spin-crossover effect, while in Ref. [24] it is shown that in $BiCoO_3$ the ferroelectric polarization is greatly enhanced when the Co^{3+} ion is in the high-spin state, as compared to the nonmagnetic state with the Co^{3+} ion in the low-spin configuration (see Table 2). They demonstrated also the predicted *electric magnetization* [18] (see point 3) above), when by means of induced polarization the spin state changes from low-spin ($S = 0$) to high-spin ($S = 2$). The authors concluded that in contrary to the widespread belief, "unpaired electron spins *actually drive ferroelectricity*, rather than inhibit it, which represents a shift in the understanding of how ferroelectricity and magnetism interact in perovskite oxides" [24].

4.3. Polar Nanoregions and Relaxor Properties

Polar nanoregions (PNR) or "polarized microdomains" in the non-polarized, paraelectric phases are observed arguably in all perovskite ferroelectrics. They are presently well studied experimentally, as they strongly influence all the properties of these systems (see [25-27] and references therein). The basic facts are: in the paraelectric phases of ferroelectrics, above the Curie temperature T_C, where the bulk crystal is cubic and no polarization is expected, small islands (nanoregions) with a limited number *n* of unit cells are spontaneously formed, their size decreasing with increasing temperature $T_n - T_C$. Above the temperature T_B, called Burns temperature, the PNR disappear, the crystal becomes regular, paraelectric. In addition, PNR disappear under external, sufficiently strong electric fields that make the crystal regularly polarized. At this stage, the paraelectric phase remains ergodic. With cooling, PNR grow in size [28], and at a temperature T_f (closer to T_C, $T_C < T_f < T_B$), relaxor properties change again to a non-ergodic, glass-like state [29-32], which then, at a somewhat smeared (diffuse) temperature around T_C, undergo the phase transition to the tetragonal polarized state. Again, distinguished from dipole glasses, this non-ergodic state of the crystal can be irreversibly transformed into a regular polarized state by strong enough external electric fields. These are *relaxor ferroelectrics*, which have a variety of properties with direct applications in materials science.

In spite of more than 60 years of research, hundreds of publications, and dramatic influence on properties, the driving force of formation of polar nanoregion (PNR) and consequent relaxor properties remained uncertain, inconclusive, although several suggestions were discussed in the literature (see in [25-27]). The majority of them, referring to the most studied mixed perovskites, relay (in different forms) on the assumption of basic structural disorder, or "random fields" produced by the differences in the active centers. However, 'random fields' or other 'raisins-in-the-cake' crystal imperfections do not explain the origin of temperature dependent size effects, in particular, the disappearance of PNR above T_B, or under polarizing electric fields. Most important, PNR are present also in pure ferroelectric perovskites ABO_3 [33-36]. Again, these considerations do not answer the question why PNR occur in ferroelectrics only, and do not show up in very similar non-ferroelectric crystals with structural phase transitions.

The briefly outlined above vibronic (PJT) theory of ferroelectricity explains also the origin of PNR and relaxor properties of ferroelectric crystals [37]. Like in all the other applications (e.g., in the extremely strong interaction with external perturbations, described in Section 5 below), the point is in the local dynamics of the dipolar distortions at the B centers, which are fully disordered (uncorrelated) in the paraelectric phase and partially ordered (correlated) in the ferroelectric phases. In bulk, at temperatures above T_C, the Helmholtz free energy of the cubic hase Φ_{cub} is lower than its value in the tetragonal phase Φ_{tetr}. Accordingly, as $\Phi = U - TS$, where U is the internal energy and S is the entropy, in the temperature interval $T_C < T < T_B$, we have:

$$T\left(S_{\text{cub}}-S_{\text{tetr}}\right)>U_{\text{cub}}-U_{\text{tetr}}. \tag{12}$$

In other words, in bulk, lowering potential energy from (average) cubic to (average) tetragonal does not provide enough energy gain, $\Delta U = U_{\text{cub}} - U_{\text{tetr}}$, to compensate the corresponding entropy loss, $T(S_{\text{cub}} - S_{\text{tetr}})$.

Yet, local polarized isles (polar nano-regions, PNR, Figure 9) with a limited, relatively small number n of centers may still be formed in a non-equilibrium self-assembly process of alignment of the local dipolar displacements [37]. As shown below, at $T_n > T_C$ the inequality (12) is compensated by the transformation of a part of the tetragonal potential energy ΔU_{tetr} of the PNR into work of formation of its surface W_n and transfer heat to the environment. The gain of energy in the formation of the n-center PNR $\Delta U_n = U_n^{(\text{cub})} - U_n^{(\text{tetr})}$ consists of two contributions: $\Delta U_n^{(\text{in})}$, the energy of the internal centers, i.e., the energy of ordering the n dipoles inside the PNR (all oriented, in average, along the polarization direction), and $\Delta U_n^{(\text{surf})}$, the energy of the centers in the surface layer (the "domain wall"), that are influenced by the neighbor disordered centers of the cubic phase (Figure 9). Denoting by n and n' the total number of centers in the PNR and in its border surface, respectively, we get the following estimates for the ordering energies based on the mentioned above mean field approximation [10, 37]: $\Delta U_n^{(\text{in})} = (n - n')\,\Delta U_0$, and $\Delta U_n^{(\text{surf})} = n'(\Delta U_0 - g)$. Here ΔU_0 is the per-unit energy gain by tetragonal ordering, shown in Figure 8, and g is the loss in this energy by the surface centers because of them being in a lower mean field, $E' < E$, of the environment (the mean field contribution from a part of their environment, by the disordered centers of the bulk cubic phase, is zero). Figure 10 illustrates this situation. A rough estimate yields $g \gtrsim (1/6)\,\Delta U_0$ [37].

The potential energy of the PNR is thus $U_n^{(\text{tetr})} = U_n^{(\text{cub})} - n\Delta U_0 + n'g$, or, by taking $U_n^{(\text{cub})} = 0$ as the energy read off, we get:

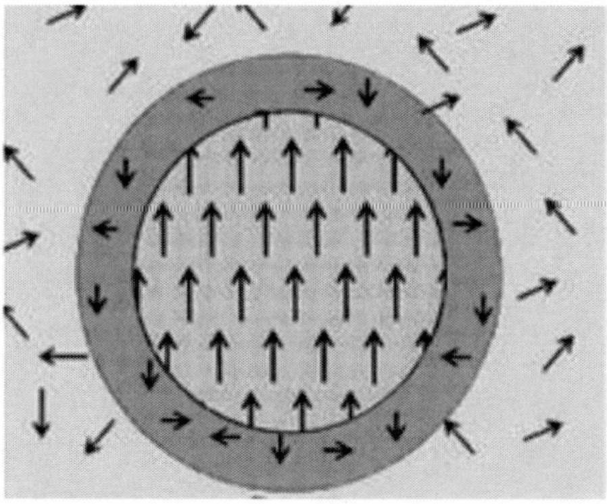

Figure 9. Polarized nanoregion inside the cubic perovskite with spherical form (center) and a border layer (dark) at temperatures T above the Curie one T_C, but below the Burns temperature T_B, $T_C < T < T_B$. Arrows indicate the direction of the local averaged dipole moments, which are tetragonally ordered inside the PNR and fully disordered in the cubic phase [37]. For the role of the PNR surface layer, see the text.

$$U_n^{(\mathrm{tetr})} = -n\Delta U_0 + n'g \tag{13}$$

Consider, for example, the PNR of spherical form (which seems to be dominant) with the radius *r*. In the cubic crystal with the unit cell dimension *a* the number of centers in a spherical PNR equals the ratio of their volumes:

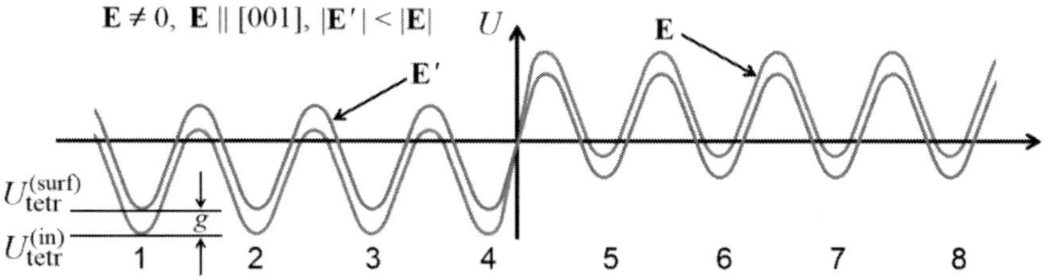

Figure 10. (Color online). Schematic cross-section of the APES $U(Q_x,Q_y,Q_z,)$ along the path of steepest descent from the orthorhombic saddle-points to the trigonal minima for the surface centers of the PNR where the tetragonal mean field (E, E′ ∥ [001]) is weaker, |E′| < |E| (purple curve, upper for 1,2,3,4, and lower for 5,6,7,8). The lowest minima on the APES are shallower than the ones for the inside centers (green curve, lower for 1,2,3,4, and upper for 5,6,7,8) by the *g* value; cf. Figure 8 [37].

$$n = \frac{\text{Volume of the sphere}}{\text{Volume of one unit cell}} = \frac{\frac{4}{3}\pi r^3}{a^3} = \frac{4\pi}{3}\left(\frac{r}{a}\right)^3. \tag{14}$$

Hence,

$$\frac{r}{a} = \sqrt[3]{\frac{3n}{4\pi}} = \sqrt[3]{\frac{3}{4\pi}} n^{1/3} \approx 0.62 n^{1/3}, \quad \text{or} \quad r = a\sqrt[3]{\frac{3}{4\pi}} n^{1/3} \approx 0.62 a n^{1/3}, \tag{15}$$

while the number of cells on the surface equals

$$n' = \frac{\frac{4}{3}\pi r^3 - \frac{4}{3}\pi (r-a)^3}{a^3} \approx 4\pi\left(\frac{r}{a}\right)^2 = \sqrt[3]{36\pi} n^{2/3} \approx 4.8 n^{2/3}. \tag{16}$$

Substituting Eqs. (9) and (11) into (8), we get:

$$U_n^{(\text{tetr})} = -\frac{4\pi}{3}\left(\frac{r}{a}\right)^3 \Delta U_0 + 4\pi\left(\frac{r}{a}\right)^2 g = -\frac{4\pi r^3}{3a^3}\Delta U_0 + 4\pi r^2 \alpha, \tag{17}$$

where $\alpha = \dfrac{g}{a^2}$.

Physical meaning of this result follows from the presentation in terms of radial force,

$$f = -\frac{\partial U}{\partial r} = -\frac{\partial}{\partial r}\left(-\frac{4\pi r^3}{3a^3}\Delta U_0 + 4\pi r^2 \alpha\right) = 4\pi r^2 \frac{\Delta U_0}{a^3} - 8\pi r \alpha. \tag{18}$$

The first, positive term describes the internal “pressure” in the radial direction tending to enlarge the PNR,

$$P = \frac{\text{Radial force}}{\text{Surface area of the sphere}} = \frac{4\pi r^2 \left(\Delta U_0 / a^3\right)}{4\pi r^2} = \frac{\Delta U_0}{a^3}. \tag{19}$$

It reflects the tendency to gain internal energy PV with $V = (4/3)\pi r^3$ by increasing the number of ordered dipoles. Noteworthy, the "pressure" $P = \Delta U_0/a^3$ = const is size-independent. Hence, growth of PNR is an "isobaric" process. The second term, originating from the surface layer, is negative; it compresses the sphere, quite similar to "surface tension":

$$\Delta P_{\rm st} \approx -\frac{8\pi R\alpha}{4\pi r^2} = -\frac{2\alpha}{r}, \qquad (20)$$

where the introduced above $\alpha = \frac{g}{a^2}$ plays the role of the coefficient of "surface tension". Notably, its negative sign in Eq. (18) means that it acts as an "external" force, and the last term in Eq. (17) describes the work, $W = 4\pi R^2\alpha$, performed on the PNR by the force of "surface tension". This interpretation, treating the first term in Eq. (17) as PV with the "pressure" (18), and the second term as work against "surface tension", is just an analogy for a better demonstration of the physics of the phenomenon. In fact, both terms follow from ordering of the PJTE induced dynamic polar displacements.

For the system in a thermostat under consideration, exchange of energy with the environment may take place without free energy conservation, $\Phi_i \neq \Phi_f$, the total energy balance being preserved by compensation of heat transfer, $Q = T(S_f - S_i) = T\Delta S$, with internal energy change ΔU and mechanical work of internal forces, $W_{\rm intern}$. Similar to other processes with heat (entropy) transfer, formation of PNR is a non-equilibrium thermodynamic process, described by Gibbs free energy change, $\Delta G = \Delta H - T\Delta S$, where ΔH is the change of enthalpy H. With growing size of PNR, their Gibbs free energy decreases and reaches a minimum value when n satisfies the condition of thermodynamic equilibrium with the environment, at which point the shape-restoring force, $-dG/dr$, is close to zero, while according to the fist law of thermodynamics, $\Delta U = Q - W_{\rm intern}$. In our case, in the process of the n-center PNR formation the system changes from the initial cubic disordered phase to the final partially ordered tetragonal one, for which $Q = T_n\left(S_n^{\rm (tetr)} - S_n^{\rm (cub)}\right)$, $\Delta U = U_n^{\rm (tetr)} - U_n^{\rm (cub)} = -n\Delta U_0$, and $W_{\text{intern}} = 4\pi r^2\alpha \approx 4.8gn^{2/3}$. Note that the work of internal forces is positive, $W_{\text{intern}} > 0$, whereas the heat transfer is negative, $Q < 0$, meaning heat is transferred to the environment. Accordingly, the first law of thermodynamics in this case acquires the following form:

$$T_n\left(S_n^{(\text{cub})} - S_n^{(\text{tetr})}\right) = n\Delta U_0 + W_{\text{intern}} = n\Delta U_0 + 4.8gn^{2/3} \tag{21}$$

Using the equation of the macroscopic phase transition $T_C\left(S_0^{(\text{cub})} - S_0^{(\text{tetr})}\right) = \Delta U_0$ and substituting $S_n^{(\text{cub})} = nS_0^{(\text{cub})}$ and $S_n^{(\text{tetr})} = nS_0^{(\text{tetr})}$, we get the relation between the temperature increment $\Delta T_n = T_n - T_C$ and the size n of the PNR, illustrated in Figure 11:

$$\Delta T_n = T_C \frac{4.8g}{\Delta U_0} n^{-1/3}, \tag{22}$$

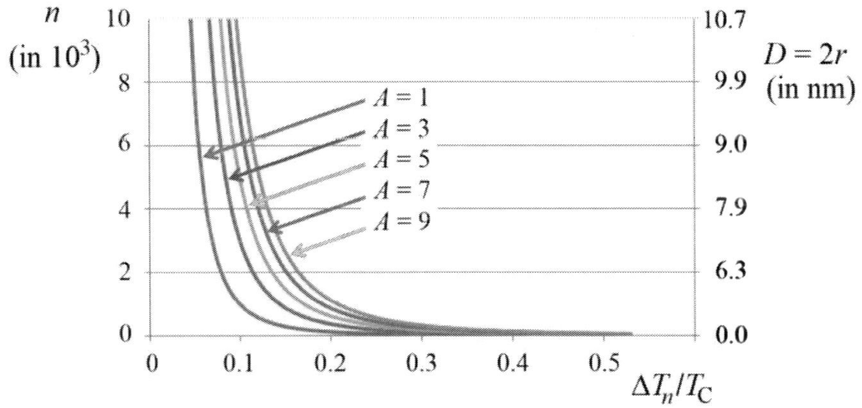

Figure 11. (Color online). Temperature dependence of the PNR size shown by the number of centers n (in 10^3, left scale), and the diameter of its spherical form D (in nm, right side scale), as a function of the crystal parameter A.

or, equivalently,

$$\frac{T_n}{T_C} = 1 + \frac{4.8g}{\Delta U_0} n^{-1/3}, \qquad T_n = T_C\left(1 + \frac{4.8g}{\Delta U_0} n^{-1/3}\right).$$

To discuss this result in more detail, following Eq. (18), we see that the force of the inner pressure is proportional to r^2 (to the square of the border surface $4\pi r^2$), whereas the compressing force $4\pi r\alpha$ is proportional to r. This means that, since $r > a$, and $n = \frac{4\pi}{3}\left(\frac{r}{a}\right)^3$, at $n \geq 4$ the force increasing the PNR size is dominant. Although the thermodynamic considerations employed in the treatment of this process are not applicable to very small numbers of centers, this conclusion shows the tendency in a qualitative manner. It means that in the fully disordered paraelectric phase, any small fluctuation that aligns the averaged dipole moments of 3-4 centers, their mean field triggers further ordering, causing the growth of the PNR. Like in a process of self-assembly, the local ordering continues until the energy gain due to the progressive loss of entropy, $(-T\Delta S)$, is exhausted by covering the loss in potential energy $n\Delta U_0$ and the work against the compressing effect of "surface tension". The entropy part is transferred to the environment (dissipated) in the form of heat, $Q_n = T\Delta S_n$. Obviously, this process cannot be reversed: the probability of a heat (or entropy) transfer from the bulk back to the given PNR is negligible. This provides stability to the PNR in the disordered environment. Once grown to its equilibrium size (to the minimum of the noted above Gibbs free energy), the PNR cannot be destroyed by thermal fluctuations unless the temperature is increased or an external electric field is applied. These estimates show also that, albeit small, an initial trigger (fluctuation) is needed to start the growth of a PNR. This explains the role of crystal imperfections or structural irregularities (particularly, in mixed perovskites) in the formation of PNR.

From Eq. (22) we get the temperature dependence of the number of centers in the thermodynamically equilibrated PNR:

$$n = A\left(\frac{T_n}{T_C} - 1\right)^{-3}, \tag{23}$$

where $A = 36\pi\left(\frac{g}{\Delta U_0}\right)^3$ is a crystal parameter. With the dependence of the number of centers n on the radius r of the spherical PNR (Eq. (14)), we obtain also its diameter D as a function of temperature,

$$D = 2r \approx 1.24a\sqrt[3]{A}\left(\frac{T_n}{T_C} - 1\right)^{-1} \quad (24)$$

With the mentioned above $g > \frac{1}{6}\Delta U_0$, we get for the crystal parameter $A > 0.5$. A reasonable estimate yields possible values $1 < A < 10$ for different crystals. For instance, for $BaTiO_3$, $T_C = 360$ K, $T_B \approx 600$ K, $A = 5$, and for $T \approx T_B \approx 600$ K we get $n \approx 17$, which for a spherical PNR means $D = 2r \approx 1.3$ nm. This is the size of the PNR, which is assumed to become undetectable in $BaTiO_3$ at Burns temperature. The temperature dependence of $n(T)$ и $D(T)$ as a function of the crystal parameter A are shown in Figure 11.

Subject to the power "–3" in Eq. (23) (see also Figure 11), the increase of the size of the PNR with decrease of temperature is very fast, especially at smaller increments $\Delta T_n/T_C \approx 0.1$-$0.2$. If, for instance, the temperature decreases linearly with time t, $T = T_{init} - mt$, where $T_{init} > T_C$, and $m =$ const > 0, then the rate of PNR growth is:

$$\frac{dn}{dt} = \frac{3mA}{T_C}\left(\frac{T}{T_C} - 1\right)^{-4} = \frac{3AmT_C^3}{\Delta T^4} \quad (25)$$

where $\Delta T = T - T_C$. For $BaTiO_3$ in the above conditions, $T_{init} \approx T_B \approx 600$ K and $A = 5$, even slow cooling with the rate $m = 0.1$ K/s yields a very fast increase of $(7 \times 10^7)\Delta T^{-4}$ centers per second. At the time when the temperature difference $\Delta T = T - T_C$ decreases to the order of 10 K, randomly polarized and arbitrarily located PNR would cover the whole crystal sample, while their reorientation and ordering require much more time. As a result, at certain temperatures T_f, $T_C < T_f < T_B$, a nonergodic “glassy” polar phase occurs (like in dipolar glasses). It follows that two conditions are necessary for the occurrence of the nonergodic phase in relaxor ferroelectrics: 1) a high concentration of PNR, and 2) relatively fast cooling. The first of these conditions is related also to structural inhomogeneity created by crystal imperfections and impurities; they favor the mentioned above local fluctuations, which trigger the PNR formation. This explains why PNR are best seen in mixed perovskite crystals. In sufficiently strong electric fields all the PNR are repolarized in the same direction, so the glassy state disappears irreversibly, distinguished from polar glasses, but in full accordance with experimental data.

5. EXTREMELY HIGH SENSITIVITY OF MATERIALS WITH JTE OR PJTE TO EXTERNAL PERTURBATIONS

As outlined above, the local distortions in the molecular and crystal centers, which are dynamic and independent at higher temperatures (before their ordering in a phase transitions), are exclusive features induced by the JTE and PJTE, which are most important also in applications that involve interactions with external perturbations. For solid state dielectrics (and similar molecular systems), very important are their properties in external electric and magnetic fields, as well as under strain and stress, in particular, permittivity, flexoelectricity, electrostriction, and structural phase transitions under pressure. Usually theoretical works on these topics are treated in the adiabatic approximation with full separation of electronic and nuclear motions, assuming that nonadiabatic effects in such systems are negligible. While in most cases correct, this assumption, as shown above, does not hold in some special, but very important situations of the JTE and PJTE. The latter, if sufficiently strong, result in spontaneous symmetry breaking that leads to two or several or an infinite number of equivalent locally-distorted equilibrium configurations with nonadiabatic (mixed electron-nuclear) states in them [1-5].

In the absence of external perturbations and sufficiently strong interactions between the JTE or PJTE centers (at temperatures above the structural phase transitions in case of solids), the distortions at each center are independent from the others, and they locally resonate between the different equivalent orientations controlled by the multiminimum APES. In the presence of a uniaxial external perturbation (e.g., strain, stress, or an electric field $\boldsymbol{E}$) the equivalency between the minima of the APES is violated, and the system becomes trapped in one of the minima (or in an orientational controlled linear combination of two or several minima that remain equivalent in the external field), in the same direction for all the JTE or PJTE centers. This results in drastically changed observable properties in terms of orders of magnitude. Below we demonstrate these effects on flexoelectricity, permittivity, and electrostriction induced by the PJTE in perovskite crystals of $BaTiO_3$ type [9, 38-40].

5.1. Flexoelectricity

Polarization of centrosymmetric dielectric solids under a strain gradient, called flexoelectricity, was first suggested long ago [41], but gained attention just recently, when it was shown that it affects significantly the properties of dielectrics, and it may have interesting applications (see in [42-52]). The physical origin of flexoelectricity is, in general, quite transparent: a strain gradient removes the inversion symmetry of the lattice in at least one direction, thus inducing a polar distortion of the charge distribution. All the theories show that in usual dielectric crystals the polarization under the flexoelectric effect is rather small, of the order of nCm^{-1}. In a recent paper [45] it was concluded that this is approximately the upper limit of the flexoelectricity strength that can be achieved in ionic crystals.

Meanwhile the experimental measurements of the flexoelectricity effect in ferroelectric systems show that it can be much larger than the theoretical predictions by 3-4 orders of magnitude [46, 47]. The largest effect is obtained in the paraelectric phase of $BaTiO_3$, but it is much smaller in the similar $SrTiO_3$ crystal, while it approximately follows the theoretical predictions in the majority of other nonferroelectric crystals. For $BaTiO_3$ the flexoelectric coefficient f (the component of the tensor coefficient of proportionality between the polarization and strain gradient in the one-dimensional model) decreases significantly when moving from the paraelectric phase to its ferroelectric phases of consequent lower symmetry and becomes "normal" (of the order of magnitude predicted by the theory) in the rhombohedral phase. In ab initio calculations for the latter $f \approx -0.37\pm0.03$ nCm^{-1}; for $SrTiO_3$ $f \approx -1.38\pm0.65$ nCm^{-1} [48].

Some of the above-cited authors stated that the origin of this big discrepancy between the theory and experiment is at present not understood, others relate it to crystal imperfections, assuming that in the paraelectric $BaTiO_3$ there are polar nanoregions (Section 4.3) that are reoriented by the strain gradient. An expert conclusion [45] states that "The magnitude of upper bonds obtained suggest that the anomalously high flexoelectricity coupling documented for perovskites ceramics can hardly be attributed to a manifestation of the static bulk effect." But in a more recent paper [49], using a variety of experimental procedures including pyroelectric current, it is shown that the effect is intrinsic (bulk), and it is due to some polar formations in dielectric solids that exist beyond (and independent of) the external strain. The authors [49] suggest also some hypothetical causes of this built-in dipolar behavior referring to procedures of the crystal (ceramic) synthesis.

The explanation of the origin of flexoelectricity in perovskite crystals is thus full of significant controversies, and we would summarize them by formulating a question, which seems to be illustrative: why the flexoelectric coefficients of two very similar perovskites crystals, paraelectric $BaTiO_3$ and $SrTiO_3$, prepared under the same conditions, differ by several orders of magnitude? Of course, there may be some differences between the paraelectric barium titanate and the very similar in atomic and electronic structure strontium titanate due to the difference in the sizes of the two ions, Ba^{2+} and Sr^{2+}, but in the usual theoretical description of ionic crystals this cannot amount to the observed several orders of magnitude difference in the flexoelectric coefficient. No one of the papers, cited above, gives a reasonable explanation of this controversy beyond the unproved assumption of different crystal imperfections.

The essential controversies in the origin of flexoelectricity disappear, at least on a qualitative level, when the JTE or PJTE are taken into account. In treating the lattice-strain interaction, the papers above do not take into account the details of the local electronic structure and the described in this paper vibronic PJTE or the JTE. These effects under certain conditions produce a local dipolar instability and dipolar distortions, which in the free system are of dynamic nature, but become oriented by the gradient of strain. As outlined in this paper, in ferroelectric crystals the broken-symmetry configurations at its centers are dipolar, their sufficiently strong interaction resulting in temperature-dependent ferroelectric and paraelectric phases, or remain weak-interacting centers with resonating, dynamically disordered dipoles. In all these cases a strain gradient restores a static orientation of the dipoles, in the same direction for all the centers, thus polarizing the centrosymmetric system. This explains the (by-orders-of-magnitude) bigger flexoelectric effect in perovskite crystals as due to the PJTE: restoring a resonating dipole moment (stopping its tunneling between equivalent minima by violating their equivalency, see below) is much easier than polarizing a rigid ionic lattice.

For a more specific treatment, we show here how the theory works for the most studied system with flexoelectric properties, the $BaTiO_3$ crystal, using the results outlined above. In the paraelectric $BaTiO_3$ the Ti ions are dipolar displaced in the trigonal minima, but in the macroscopic picture the crystal is cubic and centrosymmetric because of their noted above averaged position due to the tunneling between the eight equivalent minima (the local rotating dipoles in paraelectric $BaTiO_3$ predicted by the PJTE theory are observed experimentally, directly or indirectly (see, e.g., in [9], as well as, for instance, in tetrahertz-field-driven ionic response [50]). The data in Table 1 allow for an estimate of tunneling splitting $\Gamma \approx 35$ cm^{-1}. Any small perturbation in the form of strain gradient makes these PJTE minima nonequivalent, thus violating the conditions of tunneling between them and restoring some of the local dipole moments; the strain gradient acts as a "stop-flow" terminating a part of the tunnelings. Since the strain gradient has the same direction for all the unit cells, this leads to the polarization of the crystal.

A more rigorous approach to the problem of interactions with external perturbations is outlined below in this Section in the treatment of giant permittivity, but for a rough estimate we assume here that the tunneling is violated when the splitting of the degenerate energy level in the minima of the APES by the strain Δ_s becomes significant as compared with the tunneling splitting Γ. Because of the exponential nature of the tunneling we assume also that the restoration of the dipole moment of the unit cell *d* by the strain gradient G = dS/dx along the [111] direction follows approximately the relation $d = d_0[1- \exp(-p)]$, where d_0 is the dipole moment with the Ti ion fixed in the minimum and $p = \Delta_s/(\Gamma - \Delta_s)$, so that at $\Delta_s = 10^{-1}\Gamma$, almost the full polarization $d = 0.9d_0$ is restored (the order of magnitude of the results is not strongly affected by this approximation). Denoting the strain perturbation by W = GD, where D = dE/dG is the strain-energy coupling constant, we get for the splitting of a twofold degenerate level by strain Δ_s = 2GD. Now, the strain energy E can be estimated from the data obtained above. In polar normal displacements along [111] Q is the normalized difference between the Ti and three oxygen ion displacements, so the strain is approximately equal to $\Delta S = Q/a$, where a is the lattice constant, with the derivative $G \approx \Delta S/\Delta x = Q/a^2$. On the other hand, the gradient-strain induced energy is $E = (1/2)K_0Q^2 = (1/2)K_0a^4G^2$, wherefrom we get the strain-energy coupling constant D = dE/dG and the energy level splitting Δ_s = 2GD. Then from the equations $\Delta_s = 10^{-1}\delta_0$ and $P = f\,G$ we obtain the final formula for estimating the approximate value of f:

$$f = (d_0/a)(20K_0/\delta_0)^{1/2} \qquad (26)$$

where we used the expression for the polarization due to the dipole moments in the minima $P = d_0/a^3$. The dipolar minimum position from Table 1 is $x_0 = \sqrt{3}Q_0$ and the effective (Born) charge of the Ti ion is $Z^* = 8.7e$ [8], hence $d_0 = -\sqrt{3}Z^*eQ_0$. With numerical values of the parameters and K_0 from Table 1 we get the following estimate: $f = -\,0.43 \cdot 10^{-6}$ Cm^{-1}.

Thus instead of (and in addition to) producing a polar charge redistribution in a rigid cubic crystal of the order of nCm^{-1} the strain gradient in the paraelectric phase of a PJTE ferroelectric just recovers the virtual local dipole moments (suppresses their dynamical averaging) inducing the ferroelectric polarization of the order of μCm^{-1}. This removes the main controversy in the theory of flexoelectricity explaining the origin of the 3-4 orders of magnitude larger flexoelectric coefficient *f* in paraelectric $BaTiO_3$ as compared with nonferroelectric crystals.

The unipolar strain gradient along the [111] type direction in the paraelectric phase traps the Ti ion in one of the minima of the APES producing the polarization $P = d_0/a^3$. With the strain gradient along the orthorhombic direction two minima remain equivalent with the Ti ion resonating between them producing an effective averaged dipolar displacement of $(2/3)^{1/2}P$, approximately 0.8 of the maximum value. In the tetragonal direction the strain leaves four minima equivalent reducing the effective dipolar displacement of the Ti atom to the $(3)^{1/3} = 0.58$ part of its maximum value in the trigonal minimum. When moving from the paraelectric to the ferroelectric phases some dipolar minima become nonequivalent without the strain gradient thus eliminating enhanced flexoelectricity in some directions, and in all direction in the fully polarized rhombohedral phase.

As the JTE and the PJTE are of local origin, the enhanced flexoelectric effect takes place everywhere these effects produces local dipolar distortions, and the density of such JTE or PJTE centers is high enough for experimental observation (crystals with local dipolar distortions induced by the JTE, e.g., with tetrahedral centers in degenerate electronic states, are not explored so far). Very similar to $BaTiO_3$ are the ferroelectric properties of the $KNbO_3$ perovskite crystal ($BaTiO_3$ and $KNbO_3$ were the first two crystals in which the order-disorder nature of the phase transitions was observed by X-ray diffuse scattering [11]); the NbO_6 centers have a qualitatively similar to the TiO_6 PJTE polar instability producing local dipole moments. Whether or not these dipole moments are ordered depends on the crystal structure, more precisely, on their possible direct interaction and the ordering controlled by the environment and temperature (Section 7). This explains the origin of the enhanced flexoelectricity in perovskite systems that contain a significant amount of NbO_6 centers, e.g., $Pb(Mg_{1/3}Nb_{2/3})O_3$ [51] and $(Bi_{1.5}Zn_{0.5})(Zn_{0.5}Nb_{1.5})O_7$/Ag [52]. Obviously, many other crystals with dipolar centers of PJTE origin, including organic and biological systems, may have similar enhanced flexoelectric properties.

The $SrTiO_3$ crystal requires some additional consideration. As mentioned above, this crystal is not ferroelectric, but it is a virtual ferroelectric becoming one under negative pressure or with Ba impurities. This means that the PJTE in its Ti centers is not strong enough to obey the inequality (11) and produce dipolar instability, but its values Δ and $8F^2/K_0$ are close, so the criterion of the PJTE instability $\Delta < 8F^2/K_0$ may be induced by the strain gradient. Indeed, the latter is softening the crystal in at least one direction thus lowering the K_0 value, hence allowing dipolar distortion (polarization) in this direction. This explains (qualitatively) the order of magnitude larger flexoelectric coefficient in this crystal as compared with expected in the theory without the PJTE.

To conclude this part of the Section, strongly enhanced flexoelectricity may occur due to the JTE or PJTE, which under certain conditions produce local dipolar distortions of dynamic nature that in the average do not remove the inversion symmetry of the solid, but a strain gradient suppresses the dynamics and restores the static dipole moment, resulting in the polarization of the system. Estimates for paraelectric $BaTiO_3$ show that this effect eliminates the (more than three orders of magnitude) difference between the theoretical predictions (without the PJTE) and experimental measurements of this property.

5.2. Giant Permittivity and Electrostriction

Similar to flexoelectricity, giant permittivity of ferroelectric crystals occurs due to the dipolar distortions produced by the JTE or PJTE [39, 40] (the similarity between flexoelectricity and permittivity is well noted in the literature). In this case the equivalency between the minima of the APES is violated by an external electric field $\boldsymbol{E}$, due to which the dynamic polar distortions become trapped in one of the minima (or in an orientational controlled linear combination of the two or several minima that remain equivalent in the field ***E***), in the same direction for all the JTE or PJTE centers. In case of polar distortions this results in polarization of the system which is thus of orientational nature: the virtual dipole moment that resonates between the equivalent minima becomes real, oriented along the field. In addition to this orientational effect there is also the usual displacive polarization. The classical Debye-Curie-Weise-Langeven theory takes into account both these contributions to permittivity ε, as well as its dependence of the frequency ω of the electric field perturbation:

$$\varepsilon(\omega) = \varepsilon_0 \left(1 + \frac{\chi_d}{1+\omega^2\tau_d^2} + \frac{\chi_o}{1+\omega^2\tau_o^2} \right) \tag{27}$$

where χ_d and χ_o are, respectively, the displacive and orientational electric susceptibilities, τ_d and τ_o are the relaxation times in the corresponding relaxation processes, and ε_0 is a constant, the permittivity of the vacuum. The orientational contribution to the polarization is always much larger than the displacive one (for example, in water $\chi_o = 72$, whereas $\chi_d = 5$). This is quite understandable: similar to the case of flexoelectricity, discussed above, it is much easier to rotate a ready-made dipole than to displacive-polarize a stable polyatomic formation.

Dipolar distortions are induced by the JTE in noncentrosymmetric systems, and may occur in both centrosymmetric and noncentrosymmetric systems as a result of the PJTE [1, 3, 5]. Among the widespread cases, three (or a trough of) equivalent dipolar-distorted configurations occur in the JTE in systems with trigonal symmetry in degenerate E states (in the $E\otimes e$ problem), and four equivalent dipolar distortions take place in tetrahedral systems with degenerate T terms in the $T\otimes t_2$ problem. Among the PJTE cases, the considered in this paper $(A_{1g}+T_{1u})\otimes t_{1u}$ problem in application to the metal centers in perovskites, and its most studied $BaTiO_3$ crystals with numerical estimates of the parameters values is the best example to explore the giant permittivity [39, 40].

As noted above, in the paraelectric phase of $BaTiO_3$ the Ti ions are dipolar displaced in the trigonal minima, but in the macroscopic picture the crystal is cubic and centrosymmetric because of their noted above averaged picture due to the tunneling between the equivalent minima. In the electric field E along the trigonal axis the energy difference between two opposite minima of the APES equals $\Delta_E \cong 2d_0E$, where d_0 is the dipole moment of the unit cell when the Ti ion is locked in one of the minima. It makes the minima non-equivalent, suppressing the tunneling when it reaches the order of magnitude $\Delta_E = k\Gamma$, where Γ is the mentioned above tunneling splitting, and k is a numerical coefficient, $k \leq 1$. At the electric field $E_m \cong k\Gamma/2d_0$ all the Ti ions are displaced along the same trigonal direction, and the polarization of the crystal in this direction is $P = d_0/a^3$, where a is the unit cell constant. Hence for the orientational dielectric susceptibility χ_o in the relation $P_o = \chi_o E$, up to $E = E_m$, we have:

$$\chi_o = 2d_0^2/ka^3\Gamma \tag{28}$$

and for its contribution to permittivity we get: $\varepsilon = \varepsilon_0(1 + \chi_o)$.

The numerical data for the $BaTiO_3$ crystal in Table 1 allow us to estimate the numerical value of the susceptibility and permittivity. Taking into account that the trigonal minimum (polar normal coordinate) position with respect to the center of the cell $x_0 = \sqrt{3}Q_0$, we have: $d_0 = |e|Z^*\sqrt{3}Q_0$ where $Z^* = 7.8$ is the effective (Born) charge of the Ti ion [8]. With the numerical data for Q_0, Z^*, and δ_0 from Section 4 and in the limits of the coefficient $k = 0.1$--1.0 we get for the orientational dielectric susceptibility of the $BaTiO_3$ crystal $\chi_o = 0.45 \times 10^3$--$0.45 \times 10^4$. This agrees well by orders of magnitude with experimental measurements of permittivity in this system (see, e.g., [53]).

On the other hand, ignoring the PJTE, the displacive polarization of the unit cell P_d is limited by its stiffness (force constant) K_0 with respect to the polar distortions along the above normal coordinate Q. Hence the electric force $2|e|Z^*E$, producing the dipole moment $|e|Z^*Q$, is compensated by the elastic force K_0Q. Thus $2|e|Z^*E = K_0Q$ and $P_d = \frac{|e|Z^*Q}{a^3} = \chi_d E$, which leads us to the expression:

$$\chi_d = 2(eZ^*)^2/a^3K_0 \tag{29}$$

Using the same numerical values as above and the K_0 value from Table 1 we get the following estimate: $\chi_d = 1.22$. This yields a typical value expected for permittivity in systems with just displacive (no orientational) contribution (and in the paraelectric region far from the phase transition). Compared with the value $\chi_o = 0.45 \times 10^3$-- 0.45×10^4, obtained above for the orientational contribution, we see that *by taking into account the dynamic PJTE the calculated permittivity of* $BaTiO_3$ *increases by more than three orders of magnitude*.

In fact the whole permittivity in the field E up to the orientational saturation at $E = E_m$ is of PJTE origin, acquiring a less-significant displacive-only addition in higher fields. The value of E_m can be estimated from the above expression $E_m \cong k\Gamma/2d_0$. For $BaTiO_3$ $E_m \cong k \times 1.15 \times 10^7$ V/m. It follows that up to this voltage the permittivity is strongly enhanced by the orientational contribution of the dynamic PJTE, reaching the value of polarization $P = d_0/a^3 \cong 0.47$ Cm^{-2} which coincides approximately with that of the ferroelectric phase. Very similar to $BaTiO_3$ are the expected properties of $KNbO_3$ perovskites; as mentioned above, the NbO_6 centers have a qualitatively similar to the TiO_6 PJTE polar instability producing local dipole moments. Since the tunneling splitting is temperature dependent, the permittivity will be temperature dependent too, especially near the phase transition to the tetragonal phase (see below).

The enhanced polarization takes place in any dielectric polyatomic formation in the presence of dynamic JTE or PJTE, and the number of such systems in solid, liquid, and gas phases is innumerable. If the centers of the system with local trigonal symmetry are in a degenerate electronic E state, the local $E\otimes e$ type JTE results in three dipolar minima or a trough of dipolar distortions of the Mexican-hat type ([1-4], see also Section 3). As in the $BaTiO_3$ case, their contribution to the polarization in electric fields is expected to strongly enhance the permittivity [39]. Similarly, enhancement is expected when the centers with trigonal symmetry are subject to a PJTE.

If the electric field along the trigonal axis of $BaTiO_3$ suppresses the tunneling between the dipolar minima of the APES trapping all the units of the system in the same kind of trigonal minimum along the whole crystal, the latter will be not only polarized, but also strained in this direction. Similar to permittivity and flexoelectricity, the change in the dimensions of the unit cell in the electric field E, the electrostriction, can be estimated for the $BaTiO_3$ crystal. In fact the electrostriction problem is closely related to the inverse of flexoelectricity: instead of polarization P induced by strain gradient G in flexoelectricity, electrostriction is the strain S (not strain gradient) induced by the polarization. In the one-dimensional model with the electric field and strain along the same trigonal axis of the crystal we have: $S = \gamma P^2$, where γ is the coefficient of electrostriction. More convenient is to measure the strain as a function of the electric field intensity E, $S = \eta E^2$. As shown above, at $E_m \sim (10^6—10^7)\ Vm^{-1}$ in the trigonal direction the full orientational ordering of the local unit cell dipoles is reached resulting in the strain $S = Q/a \cong \sqrt{3}Q_0/4 \cong 0.05$. Without the orientational contribution of the dynamic PJTE, from the data above we have: $2|e|Z^*E = K_0Q$ which at $E = E_m$ yields the strain $S \sim 10^{-5}$, a typical value for a regular ionic crystal. Again, as for permittivity and flexoelectricity, we obtain a giant electrostriction of paraelectric $BaTiO_3$, about three orders of magnitude bigger than that in similar systems without the dynamic PJTE, and this effect is in accord with experimental measurements [54].

In the qualitative and semi-quantitative description above, the only requirement for the giant flexoelectricity, permittivity, and electrostriction effects to take place in matter is the presence of local centers with polar instability in several (or infinite number of) equivalent directions with free or hindered rotations of the dipoles (the case of free rotation of the dipoles is similar to a polar liquid). In the perovskites ABO_3 under consideration such polar centers may occur at B sites (in the clusters $[BO_6]$) due to the PJTE. In the more studied case of $BaTiO_3$ the polar instability is strong with an eight-minimum APES, but relatively shallow barriers between the wells. Therefore, the rotation of the local dipoles (either by tunneling or by over-the barrier transitions) is hindered. In other perovskites such a possibility may occur, subject to the local ground state being sufficiently close in energy to excited electronic states (bands) with the same spin multiplicity but opposite parity (Section 4). Similar conditions apply to any other atomic centers in matter. In all these cases free or hindered rotation of the dipoles may occur causing an enhanced response to external perturbations. The free rotation case comes about when, instead of the considered in this paper multi-minimum case, the JTE or PJTE results in a APES with a two-dimensional trough [1, 2, 4]. A more rigorous theoretical consideration of the problem is given below for permittivity of perovskite crystals ABO_3 of $BaTiO_3$ type [40]; similar treatment can be applied to the other giant effects described above.

In anisotropic crystals the linear polarization **P** induced by the electric field **E** is a tensor, $P_j = \Sigma_k \chi_{jk} E_k$. Accordingly, the permittivity is a symmetric tensor of the second rank, $\varepsilon_{jk} = \varepsilon_0 (1 + \chi_{jk})$. In cubic crystals the main three components of the tensor ε_{jk} coincide, and the permittivity is isotropic. On the other hand, in ferroelectric crystals with PJTE the orientation contribution to the susceptibility χ_o may depend on the direction of the electric field. Indeed, each elementary cell of such a crystal is unstable with respect to the displacement of the central atom along trigonal axes C_3, hence the magnitude of the vector **P** has a maximum when the electric field vector **E** points along one of these axes. It follows that in these ferroelectrics the permittivity has *cubic anisotropy* even in the cubic paraelectric phase where in the absence of the PJTE it is expected to be isotropic.

A more elaborate consideration with rigorous calculations [40] shows indeed that both the polarization and the dielectric susceptibility are angular dependent with the anisotropy dependent on the electric field intensity and temperature. Figure 12 shows the results of numerical calculations of the polarization of the crystal $P = \langle d \rangle / a^3$, with the vector $\langle d \rangle$ as the averaged dipole moment, where its dependence on both the magnitude and direction of the applied electric field E, as well as on temperature, are shown. P is given in units of $P_m = d_0/a^3$, which is the maximum value of P reached on saturation along the direction E||[111], while the magnitude of E is given in units of $E_0 = \Gamma/d_0$, where Γ, as above, is the tunneling parameter, characterizing the local dipole "orientation rate".

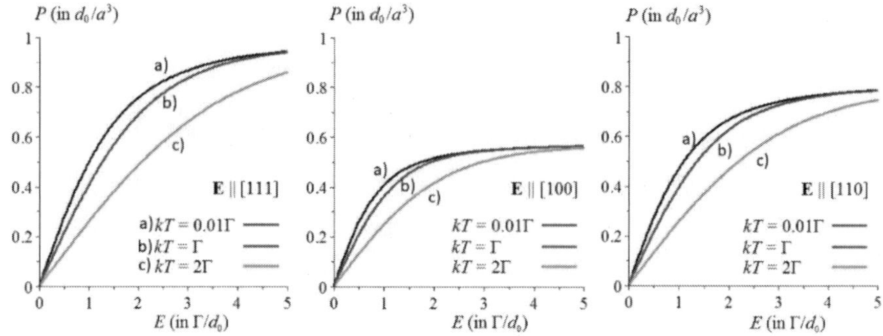

Figure 12. (Color online). Variation of the polarization P (in units of d_0/a^3) of the cluster $[BO_6]$ with the external electric field E (in units of Γ/d_0) along the main symmetry axes [111], [100] and [110] at three temperatures: a) $kT = 0.01\Gamma$, b) $kT = \Gamma$, and c) kT = 2Γ. [Reprinted with permission from Ref. [40]. Copyright 2017, Journal of Physics].

From these data it follows, first, that there is a big orientational polarization with a strong dependence on the magnitude of the electric field; no orientational polarization is expected without the PJTE. The displacive polarization of the ionic lattice is by several orders of magnitude smaller. Also, all graphs in Figure 12 show saturation of the polarization beginning from $E \approx 2\Gamma/d_0$. This means that the polarization of the crystal becomes close to saturation when the energy of the dipole moments in the electric field reaches the order of magnitude of the tunneling splitting (or over-the-barrier transitions). At this point the dipoles become fully oriented along the applied electric field. This confirms the outlined above semi-quantitative estimates of the orientational contribution to the magnitude of polarization given in [40]. Next, the differences in P values in the three directions in Figure 12, trigonal E||[111], tetragonal E||[100], and orthorhombic E||[110], show that there is a significant angular dependence, which is not expected in the absence of the PJTE. Induced by the latter, the polarization is maximal along the trigonal axes, where the absolute minima are positioned, while a smaller

P value emerges along the highest tetragonal saddle-points of the APES. The temperature and field diagrams of the polarization P are shown in Figure 13.

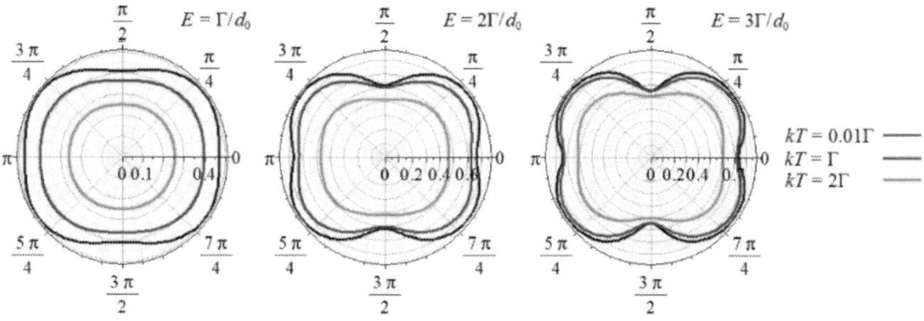

Figure 13. (Color online). Angular dependence of the polarization P in the cross-section plane $q_x = q_y$, $\varphi = \pi/4$ (see figure 1) at different electric fields *E* and temperatures: $kT = 2\Gamma$, Γ, 0.01Γ for the inner and next two curves, respectively. The polarization *P* is maximal along trigonal directions $|q_x| = |q_y| = |q_z|$, lower in the orthorhombic, and minimal in the tetragonal directions [40]. [Reprinted with permission from Ref. [40]. Copyright 2017, Journal of Physics].

A qualitative explanation of this orientation effect is given in Ref. [40]. In sufficiently weak fields $E \ll \Gamma/d_0$, all the trigonal minima remain approximately equivalent. In a stronger field, for instance, when $E \gg \Gamma/d_0$ at E||[111], the energy of the trigonal minimum with the coordinates (Q_o, Q_o, Q_o) is lowered, whereas the energy of the one at ($-Q_o$, $-Q_o$, $-Q_o$) is elevated, and there are similar differences between the other minima, which manifest pairwise correlation along, and opposite to, the direction of the applied electric field. As a result the nuclear motion is ultimately localized in the lowest well, at which point the orientation contribution to the dielectric susceptibility χ_o is frozen, and only a small contribution of the displacive one is left.

The cubic anisotropy of the polarization is due to the different magnitude of the average dipole moment at different direction of the vector **E.** For **E** || [111] in saturation all the dipoles are frozen in the lowest trigonal well along the field with the dipole moment d_0 of the unit cell. In the case when **E** || [110], two of the eight wells are lowest in energy. As they remain symmetry-equivalent in the field **E** || [110] the resonance conditions for tunneling are not violated, and the tunneling between them results in an average dipole moment of $(\sqrt{2/3})d_0$. If the applied electric field points along the tetragonal axes, **E** || [100], four wells remain equivalent in the field along the tetragonal axes, resulting in the average dipole moment equal to $(1/\sqrt{3})d_0$. The calculated values of P in figure 12 fully agree with these qualitative predictions.

The variation of dielectric susceptibility χ with the magnitude of the applied electric field pointing in three directions, E || [100], E || [110] and E || [111], and their diagrams of its angular dependence [40] are qualitatively the same as for the polarization P in figure 13. These specific properties of permittivity follow from the eight-minimum shape of the APES induced by the local PJTE and the dynamics of localization of the nuclear motion in the lowest wells situated along the applied electric field. The angular dependence of the dielectric susceptibility can be modeled by the first terms in the expansion in cubic harmonics,

$$\chi(E) \approx \chi_0(E) + \chi_4(E) Y_4(\theta,\varphi), \tag{30}$$

where χ_0 is the isotropic contribution, χ_4 is the anisotropic amplitude, and $Y_4(\theta, \varphi)$ is the fourth order cubic harmonic, $Y_4(\theta, \varphi) = (x^4 + y^4 + z^4 - \frac{3}{5} r^4)/r^4$. By fitting the numerical values of the anisotropy for $kT = \Gamma$ we find [40]:

$$\chi_0 \approx 0.89 - 0.43\frac{Ed}{\Gamma} + 0.058\left(\frac{Ed}{\Gamma}\right)^2, \quad \chi_4 \approx 0.017 - 0.39\frac{Ed}{\Gamma} + 0.102\left(\frac{Ed}{\Gamma}\right)^2, \tag{31}$$

where $E = |\mathbf{E}|$, $d = |\mathbf{d}|$.

The giant permittivity induced by the local PJTE in the B centers of the ABO_3 perovskites is evaluated above as a function of the ratio Γ/d_0, where Γ is the tunneling splitting parameter, and d_0 is the dipole moment in the minimum of the APES. The main parameter is the d_0 value that characterizes the dipolar distortion, and hence the magnitude of the observable effects. As for Γ, its main function is to stand for the speed of transitions between the minima, which may vary significantly depending on the height of the energy barrier between them. With low barriers these transitions become more of over-the-barrier nature, and at a zero barrier they are free rotations along the PJTE-induced two-dimensional trough. Qualitatively, the effects described above will be present in all such cases, subject to the dipolar instability $d_0 > 0$, but the quantitative description may be different. The "rotation frequency" of the dipoles influences mostly the magnitude of the field at which the orientation polarization is saturated.

6. SPECIAL MAGNETIC-DIELECTRIC BISTABILITY IN MATERIALS WITH THE JTE AND PJTE

Molecular systems and solids with two coexisting stable and close-in-energy configurations that exhibit markedly different magnetic and structural properties with controllable conversion between them (bistability) are very attractive for application in electronics as high-density information storage materials and sensor devices [55-58]. With regard to molecular magnetic bistability the most studied systems so far, apart from the single-molecule magnets [59], involve some specific spin crossover in transition metal complexes (TMC), for which the low-spin (LS) to high-spin (HS) transition can be induced by temperature or pressure changes, or by light irradiation [21, 57, 58]. However, in spite of more than four decades of study of such systems, the room-temperature lifetime of their metastable state in isolated complexes is typically a few nanoseconds, which is too short for practical application. Cooperative effects [60, 61] between a few thousands of these complexes may extend their lifetime at the cost of substantially increasing the size of the material.

On the other hand, it was shown [62] that in molecular systems and local centers in crystals with at least one threefold (or higher) symmetry axis and a *half-filled degenerate valence shell*, a similar to the spin-crossover (but richer) bistability effect takes place. Each of the two-fold degenerate e and three-fold degenerate t valence orbitals can be populated by two electrons with opposite spins, a total of 4 and 6 electrons, respectively. As stated above (Section 2), the physical essence of the JTE is that if there is a degenerate state with two or more charge distributions that have the same energy, each of them is necessarily not totally symmetric and distorts the system. The non-totally-symmetric charge distribution in any of the degenerate states is complemented by those of the other states, so that if all of them are proportionally populated, each with one electron, the summed closed-shell distribution is totally symmetric. Obviously, *half-filled shells* (HFS) with proportional population of the orbital states (not the spin states) have totally symmetric charge distributions too. For two-fold e and three-fold t degeneracies the closed-shell electronic distributions are (in conventional notations; arrows denote the spin state): $e^4 = (e_\theta \uparrow; e_\theta \downarrow; e_\varepsilon \uparrow; e_\varepsilon \downarrow)$ and $t^6 = (t_x \uparrow; t_x \downarrow; t_y \uparrow; t_y \downarrow; t_z \uparrow; t_z \downarrow)$, while for half-filled-closed-shells they are $e^2 = (e_\theta \uparrow; e_\varepsilon \uparrow)$ and $t^3 = (t_x \uparrow; t_y \uparrow; t_z \uparrow)$, respectively. Hence for systems with these electronic configurations, which yield totally-symmetric charge distribution, JTE distortions, at first sight, are not expected [62].

It came out that systems with the totally symmetric charge distributions of the HFS configurations e^2 and t^3 have no JTE *in all their terms*, ground and excited, formed by the interelectron interaction, including degenerate terms [62, 63] (this is a very rare case when degenerate electronic states are not JT active). Notably, in these cases there is no PJTE too, because their low-lying excited states have different spin-multiplicity, as compared with the ground state (Figure 14), and the PJTE does not mix different spin states. However, for a series of typical systems with no apparent degeneracies or pseudo-degeneracies, it was shown that a stronger involvement of the excited states may lead to the so-called *"hidden" JTE and PJTE*; they produce an additional global minimum of the APES with lower energy, which is thus the distorted ground state [62, 64]. The distortions of the HFS electronic configurations in these cases take place as a result of a strong PJTE that mixes two excited terms, as shown in Figure 14.

Analysing the charge distribution in the distorted global minima induced by the hidden JTE and PJTE, it was shown that the distortions are accompanied by *orbital disproportionation*.

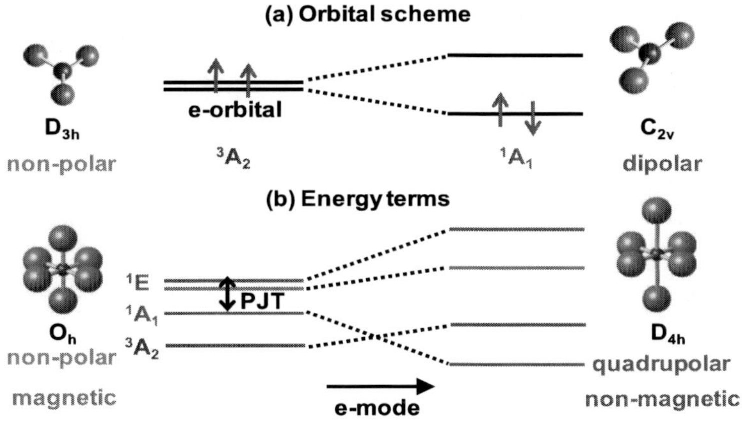

Figure 14. (Colour online). Typical energy level scheme illustrating orbital disproportionation and the hidden PJTE in systems with electronic orbital e^2 configurations: (a) in an orbital presentation with the two degenerate states proportionally populated, each by one electron, the triplet $^3A_2$ state has a totally symmetric charge distribution that does not produce any distortions of the system, but a disproportional distribution with both electrons in one of the orbitals (yielding the nonmagnetic $^1A_1$ ground state) distorts the system because of its lower-symmetry charge distribution; (b) in the (produced by interelectron interaction) energy term's presentation, with the shown typical term arrangement, there is a strong PJT coupling between the two excited states $^1A_1$and ^1E which pushes down the $^1A_1$ state to cross the $^3A_2$ state and produce another global ground state with distorted configuration [62].

This term means that instead of proportional population of all the orbitals with one electron per orbital required by the Hund's rule, the system prefers to violate the latter by occupying one of the orbitals with two electrons, thus leading to lower symmetry charge distribution and distortion. In case of e^2 and t^3 configurations, the disproportionation means electronic-configuration transitions of the type $(e_\theta\uparrow;e_\varepsilon\uparrow)\rightarrow(e_\theta\uparrow;e_\theta\downarrow)$ and $(t_x\uparrow;t_y\uparrow;t_z\uparrow)\rightarrow(t_x\uparrow;t_x\downarrow;t_z\uparrow)$, respectively, for which *the spin state changes*. In other words, the distortion of the system, produced by the hidden PJTE, is accompanied by a special *spin-crossover effect*: the undistorted (Hund) state has the higher-spin (HS) (the HS states are ^3A for e^2 and ^4A for t^3), while in the distorted configuration the spin is lower (the low-spin (LS) states are ^1A in e^2 and ^2A in t^3). Thus we come to the situation with two coexisting close-in-energy states of the molecule, high-spin undistorted and low-spin distorted, with a spin-crossover in between (that forms a barrier between these two states). More often than not the distorted configuration has a dipole moment, but no magnetic moment (spin zero), while the undistorted configuration with nonzero magnetic moment (nonzero spin) has no dipole moment, resulting in *interrelated magnetic and dielectric bistabilities* [65].

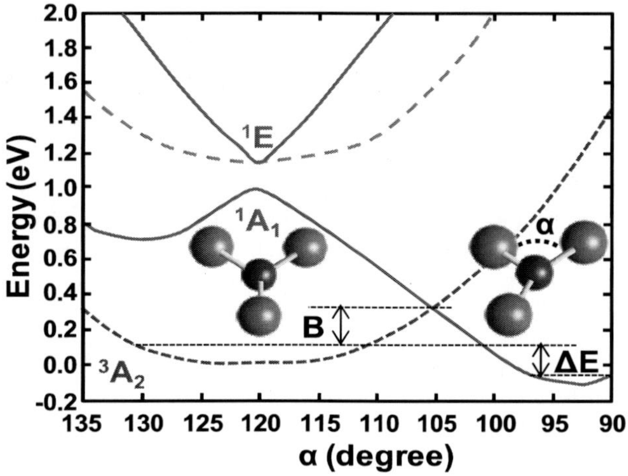

Figure 15. (Color online). Ab initio calculated energy levels of CuF_3 as a function of the angle α (*e* mode distortion) showing the formation of the two equilibrium configurations typical for HFS systems (cf. Figure 14) [65].

This picture was confirmed by ab initio numerical calculations performed on a variety of systems, including Si_3, Si_3C, CuF_3, Na_3, Na_4, Si_4, Na_4^-, C_{60}^{3-}, as well as for crystal structures of the type LiCuO2, NaCuO2 [65]. Analytical expressions were also obtained for the onset of disproportionation, showing how the gain of energy by PJTE distortion (meaning by structural changings producing better covalency) compensates for the increase of interelectron repulsion in the Hund states with higher spin [62, 65]. As the number of molecular and solid state systems with e^2 and t^3 electronic configurations is practically infinite, and they may occur in all aggregate states of matter, these finding suggest an inexhaustible source of systems with potential bistability properties.

The common properties of HFS systems and their origin can be illustrated using the CuF_3 molecule as an example. In the high-symmetry configuration this molecule is equilateral planar-triangular with D_{3h} symmetry, and its electron configuration is e^2. Figure 15 shows the results of ab initio calculations of its ground and excited electronic states in the planar configuration, as a function of the angle α(FCuF). We see that when the molecule is an equilateral triangle ($\alpha = 120^0$) the ground state is $^3A_2$ with the lowest excited states $^1A_1$ and ^1E, which is typical for all electronic e^2 configurations (see Figure 14). The spin-triplet ground state $^3A_2$ in this D_{3h} geometry is stable, but there is a strong PJT coupling between the two excited states $^1A_1$ and ^1E by *e* type distortions, due to which the lower state $^1A_1$ goes down, crosses the $^3A_2$ state, and produces a global minimum, in which the initial geometry D_{3h} is reduced to C_{2v}, and the electronic state is a spin-singlet $^1A_1$. As a result the molecule CuF_3 has two stable states: the triplet HS state has a magnetic moment of ~1:82 MB and a zero dipole moment $d = 0$, whereas the LS state has a zero magnetic moment and a strong dipole moment of $d \cong$ 1:67 D. The HS state is higher in energy by ΔE~ 0:2 eV, and it has an energy barrier of B ~ 0:33 eV to the point of spin crossover with the LS state (energy values are read off the zero vibrations).

We have thus two coexisting bistabilities, magnetic and dielectric (dipolar), which are strongly correlated: you may have either a HS state with a zero dipole moment, or a zero spin state with nonzero dipole moment. Such correlations between magnetic and polar properties are typical for solid state multiferroics. Hence the CuF_3 molecule may be termed as a *single-molecule multiferroic*. An ensemble of such molecules, e.g., organized as a thin film is a representative of HFS materials. With qualitatively the same picture of bistability different HFS systems have varying parameter values [65].

Recently [66] this molecular bistability effect has been observed directly in isolated molecules in solution containing a Ni^{2+} ion in the square-planar environment of the Ni-tetrakis (pentafluorophenyl) porphyrin complex, which has the features of the electronic e^2 configuration with energy terms of Figure 14, being thus a HFS system representative. This molecule displays two magnetic states with a low spin (S = 0) in the square-planar configuration, and high spin (S = 1) in the distorted pentacoordinated geometry, respectively, in accordance with the general conclusions outlined above. The two states are switchable by optical irradiation, and they have 27-hour half-lifetimes at room temperature. The authors [66] do not discuss the origin of their experimental results.

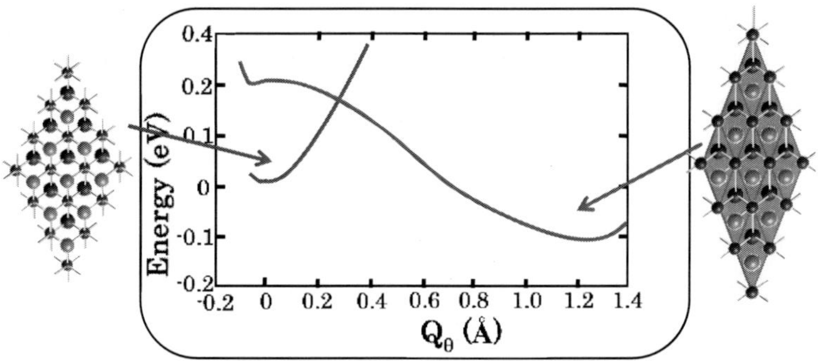

Figure 16. (Colour online). DFT calculation of the energies of the two configurations of the $LiCuO_2$ crystal, magnetic-undistorted (left curve in red) and nonmagnetic-distorted (blue), arising from the local electronic e^2 configuration of the Cu^{3+} ions in the octahedral environment of the oxygen ions, as a function of symmetrized local distortion coordinate Q_ϑ [65]. The magnetic centers are ordered anti-ferromagnetically in the crystal, while in the lower in energy nonmagnetic phase the crystal is strongly tetragonally distorted. In the two lateral illustrations, the two fragments of the layers of CuO_6 octahedrons in the corresponding two phases are shown with Cu in red, and O in light green above the plane and dark green beyond the plan.

The idea of PJTE bistabilities has been extended to solid-state systems in which case the HFS, e^2 or t^3, are formed by either localized orbitals on the centers or corresponding band states [65]. Solid-state HFS systems may be more important in applications because of their larger variety and better parameters. They may include virtually all the trigonal and cubic symmetry coordination compounds, ionic crystals, and impurity centers in crystals containing transition metal ions with the electronic configuration e^2 (d^8 ions in octahedral coordination and d^2 ions in trigonal, tetrahedral, or cubic environment), and t^3 (d^3 in octahedral and d^7 in tetrahedral coordination). Similar to the mentioned above bistability of a $Ni^{2+}(d^8)$ complex in solution [66], there is experimental evidence of a reversible equilibrium between octahedral magnetic and square-planar diamagnetic complexes of this ion in solid states [67, 68].

Electronic structure calculations were carried out for two solids from the HFS class, the crystals $LiCuO_2$ and $NaCuO_2$. Their lattices contain alternating layers of vertex-sharing $Cu^{3+}(O^{2-})_6$ octahedrons and Li^+ or Na^+ ions, respectively (Figure 16). The Cu^{3+} ion has a $d^8(e^2)$ electronic configuration (as above, *e* denotes either molecular orbitals or symmetrized band states). The results of density functional theory (DFT) calculations employing the hybrid B1WC functional [69] with geometry optimization of the two crystal structures in both the HS (ferromagnetic and antiferromagnetic) and LS (diamagnetic) states are shown in Figure 16. As predicted by the general theory and shown above for molecular cases, the HS configuration of these crystals is undistorted, while the LS state is diamagnetic with a very strong elongation of the CuO_6 octahedrons. The ferromagnetic and antiferromagnetic phases have very similar local geometries with regular octahedral coordination of the oxygen atoms around Cu, whereas the diamagnetic phase is strongly distorted to almost square-planar coordination: in $LiCuO_2$ the metal-ligand distance in the magnetic phase is 2.01 Å, while in the diamagnetic phase the equatorial and axial ligands are at 1.86 Å and 2.76 Å, respectively. In $NaCuO_2$ the same pattern is found, although the lattice parameter and distortions are by about 10% larger than in LiCuO2. The results obtained for the geometry of the diamagnetic phase of these crystals agree well (within 2%) with the experimental data [70, 71].

In view of the close energy values of the two phases, the possibility to further tune the system by means of epitaxial strain was also explored. The results are illustrated in Figure 17, where the influence of external strain is simulated by an in-plain lattice parameter. Under the influence of the strain, the two phases become almost degenerate, and hence more vulnerable to external perturbations and magneto-elastic effects [72, 73]. One of the most important characteristics of the bistability with regard to possible applications is the lifetime (relaxation time) of the induced higher-in-energy state (in most cases the spin-triplet state) with respect to its thermal relaxation to the lower (spin-singlet) state.

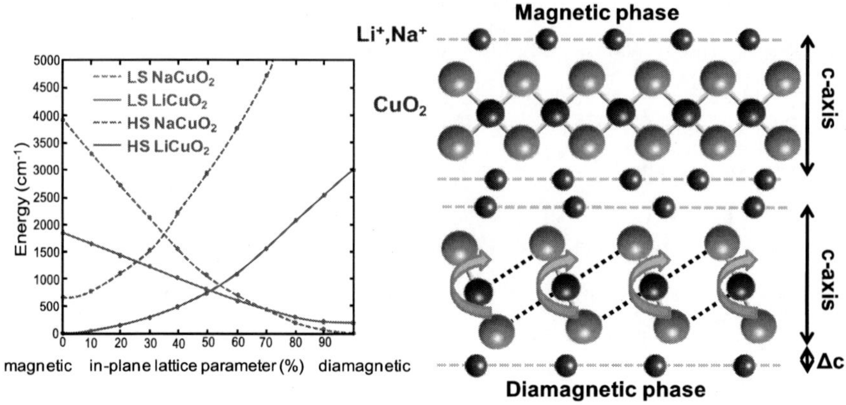

Figure 17. (Color online). Calculated relative stability of the LiCuO2 and NaCuO2 crystals in the magnetic to diamagnetic transition as a function of the in-plane lattice parameter: at 35% of strain in $NaCuO_2$ and 50% in $LiCuO_2$ the two phases become approximately degenerate. The distortion along the c axis is also illustrated [65].

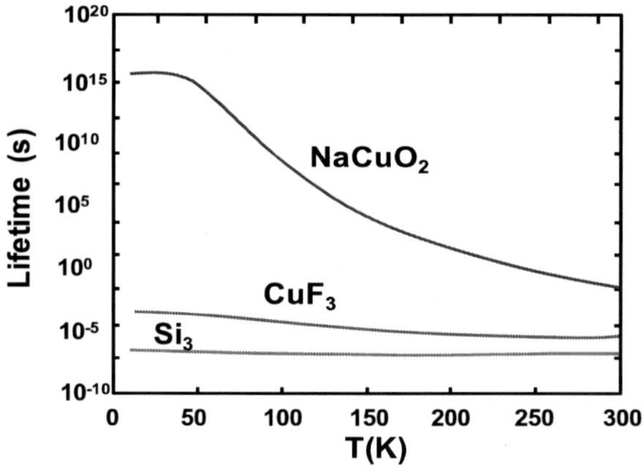

Figure 18. (Color online). Temperature dependence of lifetimes estimated for the higher in energy high-spin states of Si_3, CuF_3 and an isolated octahedral center in $NaCuO_2$ [65] by means of model calculations following Ref. [58].

This is less a problem for solids, where the cooperative interactions stabilize well the two states in the bistability. For molecular systems or small aggregates, the relaxation lifetime depends on several factors, mainly on the spin-orbital interaction and the barrier height and width between the two minima of the APES, and the dissipation of energy in interaction with the environment. Direct calculation of these lifetimes is not trivial; the estimated approximate values of the HS configuration lifetime and their temperature dependence for some molecular systems and for (isolated) octahedral units in $LiCuO_2$ are illustrated in Figure 18 [65]. The structural changes added to the magnetic ones increase essentially the lifetime of HFS systems, because of the strong dependence of the coupling constant on the overlap of the vibrational functions of the two states. Along this line, the system with larger lifetime is the CuO_6 complex in $LiCuO_2$ due to its large distortion (Q_9 ~ 1.24Å), followed by Ge_4 and Si_4 clusters (Q_9 ~0:8 Å). An increase in lifetime can also be achieved in organic and other compounds of non-heavy elements due to their smaller spin-orbital coupling constant that mixes the two states.

The calculations show that Si_3 has the shortest lifetime, which increases in systems with larger distortions and reaches macroscopic values in Si_4, Ge_4, and $LiCuO_2$ at low temperatures (but in Ge_4 there are low-lying triplet excited states that may reduce the lifetime) [65]. In the crystals $LiCuO_2$ and $NaCuO_2$ the estimates of the lifetime of the HS configuration are carried out for a separate cluster unit CuO_6; it is extremely large at low temperatures (due to the large distortions in the diamagnetic phase) and remains macroscopic at room temperature. Since the cooperative interactions between the complexes in the condensed phases obviously increase the lifetimes, these values show the lower limit of the expected lifetimes. This means that (at least from the theoretical point of view under consideration) the crystals $LiCuO_2$ and $NaCuO_2$ are quite feasible for practical use as bistability materials, and their large lifetime of both states in this bistability indicates that they can be used in the form of clusters, quantum dots, or thin films.

As mentioned above, a somewhat similar spin-crossover phenomenon is known to take place in transition metal compounds (TMC) with d^4-d^7 electronic configurations and specific values of the ligand field, for which the LS (realized in sufficiently strong ligand fields) and HS (for weak fields) states are close in energy [60, 61], so that for some ligand fields they cross over. In spite of three decades of intensive study (in principle, such two spin-states may serve as molecular materials for electronics [61]), so far the two states have been observed only for some compounds in optical LS→HS excitation at low temperatures < 50 K, and mostly as a cooperative effect in solids; the room-temperature lifetime of their metastable state in isolated complexes is typically a few nanoseconds [57, 58], which is too short for practical application. Cooperative effects [60, 61] between a few thousands of these complexes may extend their lifetime at the cost of substantially increasing the size of the material.

The PJT-induced spin crossover, outlined above, is quite different in nature than the TMC one, and takes place in a variety of molecular systems, small to moderate, organic and inorganic, as well as in metal-containing molecules. It is important that the two spin states of the HFS systems have different nuclear configurations, distorted and undistorted, with 3–4 times larger differences in interatomic distances, than in TMC systems. Therefore, the HS—LS relaxation rate in the PJT-induced case is much lower than in TMC. In addition, the spin-orbital interaction in light-atom molecules is smaller by orders of magnitude than in TMC. Remarkably, for the d^n electronic configurations that result in HFS e^2 and t^3 with the spin-crossover effect, described above, there is no ligand field induced spin-crossover.

More details about the properties induced by the PJTE in the two crystals, LiCuO2 and NaCuO2, were revealed, in particular, a giant magneto-elastic effect [65].

7. Manipulation of Structure and Properties of Two-Dimensional Systems Employing the PJTE

7.1. Introduction to the Problem

Two-dimensional (2D) or quasi-2D systems have unique properties due to a variety of their planar-surface-specific physical and chemical properties. The best-known example of this kind is graphene [74]. Similar properties are sought for in silicene, germanene, stanene, boron nitride, gallium nitride, phosphorene, graphene oxide, zinc oxide, 2D transition metal dichalcogenids, and there is an unlimited number of other organic and inorganic 2D and quasi-2D compounds with expected similar properties (see recent reviews [75-80]). Many of these systems have excellent electronic and photonic properties, exceptional mechanical flexibility, and partial optical transparency, as well as well-defined crystalline structure. The variety of their electronic structures includes isolators (e.g., hexagonal boron nitride), semiconductors (e.g., transition metal dichalcogenids), and semimetals (e.g., black phosphorene). The importance of such systems stems from their impressive, already realized and virtual applications in physical and chemical materials devices (see in [75-80]).

One of the main structural features that make these compounds special in applications is their planarity or deviation from planarity via puckering or buckling of their native (assumed planar) 2D structure. This buckling feature influences all the properties of the system, notably the band gap and charge mobility. As stated by a group of experts [81], "the native buckling can have dramatic implications in the overall physics and chemistry" of 2D or pseudo 2D systems, and "potentially serve as a nanotechnology platform ... for fundamental research and applications as reconfigurable devices for electronic, spintronic, photonic, thermal, energy, mechanical, chemical, and sensor nanosystems" and "The degree of buckling can dictate the electronic character or influence the chemical reactivity. This highlights the concept of a pseudo-planar flatland, where buckling can foster a number of potential functionalities that can be readily engineered" [81].

The planarity and deviation from planarity of 2D systems is thus one of the most important, sometimes crucial, properties in defining their possible applications. This raises the problem of influencing (manipulating) the planarity and related properties of quasi-2D systems by means of external implications (perturbations). Such possibilities were tried recently by influencing 2D systems with 3D (or nD) ones in a layer-structure combination, were they interact via van-der-Waals forces [77]. Some authors suggest influencing the electronic structure "by changing the group IVA element, the degree of spin–orbit coupling, the functionalization chemistry or the substrate" [81]. Other possibilities were also suggested [75, 82]. For instance, band gaps in atomically thin 2D semiconductor alloys can be tuned by means of controlled growth of composition, where the band gap depends on the number of layers [83].

Many attempts to influence the 2D planarity employ the considerations of atomic orbital hybridization. Indeed, compared with the ideally flat graphene, the interatomic distance in the series of buckled hexagonal honeycomb quasi-2D structures of Si, Ge, and Sn increases significantly from left to right (1.42 Å in graphene, 2.28 Å in silicene, etc.). This weakens the overlap between their p_z orbitals and reduces the contribution of *p*-bonding in the flat configuration, making the atomic sp^3 hybridization and the consequent buckled (honeycomb) structure advantageous. Note, however, that this qualitatively correct consideration of the influence of atomic orbital hybridization on local structural elements may be insufficient to explain the origin of buckling in more complicated situations. On the other hand, it was shown that all deviations from high-symmetry configurations of polyatomic systems in nondegenerate states (in our case the puckering or buckling the planar configurations) are due to, and only to the pseudo Jahn-Teller effect (PJTE) [1-5] (in degenerate states, both the JTE and PJTE may be active). In application to a series of cyclic and heterocyclic compounds the PJTE explains the origin of their out-of-plan distortions (see [3] and references therein; puckering as a PJTE was first formulated in Ref. [84]).

The knowledge of the PJTE origin and mechanism of out-of-plan distortions of 2D systems leads directly to the possibility to change their planarity by means of external influence that changes the parameters of the PJTE. In the simplified two-level formulation (see below) the condition of instability of the high-symmetry configuration of the system is a relation between its main three parameters, shown above in Eq. (11)), $\Delta < 8F^2/K_0$, where, as above, Δ is the energy gap to the excited state with appropriate symmetry (active in the PJTE), F is the constant of vibronic coupling to the latter, and K_0 is the primary force constant, characterizing the stiffness of the system with respect to the distortion.

The procedureof influencing the planarity of the 2D system, described below, starts with electronic structure calculations of the ground and low-lying excited states in the high-symmetry (planar) configuration, revealing the active excited states which produce the instability of the ground state, their energy, symmetry, molecular orbital (MO) origin, and the numerical values of the parameters in the condition of instability (1). This gives us the necessary insight on how we can influence (enhance or suppress) the condition of instability of the planar configuration, meaning manipulate the puckering or buckling of the 2D system. In principle, there are several possibilities to influence the PJTE distortions, among which the following seem to be most practical: (1) increasing the energy gap Δ between the electronic states (that are mixed by the PJT vibronic coupling) by means of weak coordination with outside atoms or atomic groups (e.g., by accompanying the 2D layer under consideration with intercalated other atoms or layer with appropriate properties), (2) changing the positions and symmetry of the PJT active ground and excited states via a redox process, by removing or adding electrons, and (3) modifying the system by means of substitutions which do not violate the quasi- 2D character of the system but change the PJTE parameters. All these possibilities were effectively realized in changing the planarity of molecular systems as shown below [85-103].

Unfortunately, the role of the PJTE in the planarity of 2D systems, and especially the possibility to control their puckering or buckling via influencing the PJTE parameters, remains unknown to the majority of the researchers of this important topic (no mention of this possibility in the latest reviews [74-83]). Together with the recent papers [6], the present section is deemed to fulfil this gap. The intention is to show how the briefly introduced above PJTE can be employed to control the puckering or buckling of the planar 2D systems using the already published papers on this topic as illustrative examples. In the section below "Revealing the excited states and PJTE parameters that produce the planar instability of 2D systems" we show briefly the technical procedure in application of the PJTE theory to the specific problem of the origin of instability of planar configurations of molecular systems. Then the Section "Flattening puckered (buckled) 2D and quasi 2D systems by suppressing the PJTE" reviews the publications in which the methods of suppressing the PJTE in puckered (buckled) molecular and some extended 2D systems and thus restoring their planar configuration, are realized.

A notice is due about a problem so far remaining unsolved: the PJTE theory and its applications are worked out in detail for finite size molecular systems, and it was extended to dielectric crystal structures with local (independent) PJTE instabilities [104]. So far there is no general theory of "band-PJTE", where the interacting ground and excited electronic states under low-symmetry phonon displacements are essentially band states, and the whole effect is summarized over the band structures. A similar problem for linear systems with respect to specific doubling-period inline distortions was solved by Peierls [105], and it was shown that it is in fact a sum of Jahn-Teller and pseudo Jahn-Teller interactions between the band states (see in [1, 4]). The band-PJTE can be viewed as an unsolved Peierls-like problem for 2D systems (instead of 1D ones) and low-symmetry 3D phonons (instead of the inline 1D ones).

Note, however, that the band presentation of the problem does not diminish the fact that the driving force of the PJTE instability and distortions is of local origin, induced by better covalency in the bindings between the atoms in the distorted configuration. Therefore, although the solution of the problem of planar instability in a band-PJTE presentation may yield additional information about the parameter values, the qualitative results obtained for the basic structural units remain valid for the whole 2D system, especially with regard to the vulnerable parameters of the PJTE to be targeted in manipulation of the 2D structure and properties. In many cases, the local instabilities and distortions remain almost unchanged in the overall 2D structure (see below, e.g., black and blue phosphorene illustrated in Figure 27).

7.2. Revealing the Excited States and PJTE Parameters that Produce the Planar Instability of 2D Systems

Based on the general proof that all the deviations from high-symmetry configuration of polyatomic systems in nondegenerate states are of PJTE origin, we can find out the main features of the planar configuration of 2D systems that are responsible for the out-of-plan instability by means of electronic structure calculations including vibronic coupling. In view of further applications (manipulations), the PJTE parameters of interest include the most active excited states responsible for the out-of-plan instability of the ground state, the energy gap to them, the symmetrized atomic displacements along the coordinate of instability, and the energy profiles of these electronic states along this coordinate, as well as the vibronic coupling parameters characterizing the strength of instability.

Below in this section we demonstrate the methodology in the procedure of evaluating these parameters using specific molecular systems as examples. First we employ a series of four 2D-hexa-heterocycles, 1,4-$C_4N_2H_4X_2$ (X = H, F, Cl, Br), where the problem can be reduced to a two-level one [85]. Starting with the planar configuration, ab initio calculations of the electronic structure and vibrational frequencies of the system were carried out. All the four compounds have a nondegenerate ground electronic states A_{1g} and an imaginary frequency of b_{3u} symmetry with the values (in cm^{-1}) 591.0, 554.5, 381.0, and 392.2, for X = H, F, Cl, Br, respectively. This means that the planar configuration of these compounds is unstable with regard to the symmetrized b_{3u} displacements of its atoms (the calculations [85] give also the directions and relative values of these displacements). According to the PJTE the excited states that produce this instability of the ground state should have the B_{3u} symmetry. If we can assume that the main contribution to the instability comes from only one low-lying excited state of B_{3u} symmetry, then the PJTE problem is simplified to a two-level one, $(A_{1g} + B_{3u})\otimes b_{3u}$. It is illustrated in Figure 19.

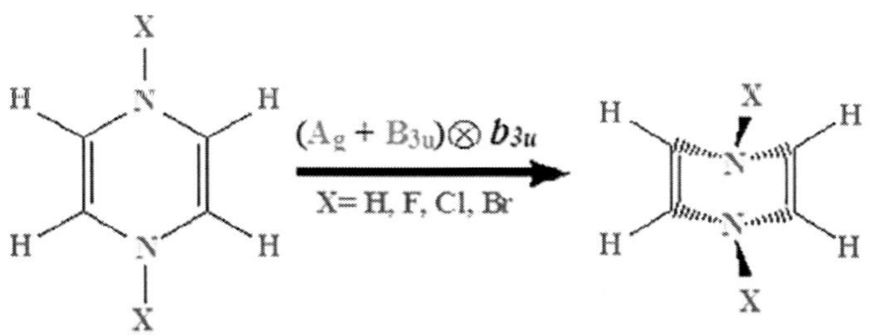

Figure 19. (Color online). Schematic presentation of the b_{3u} type out-of-plan distortion produced by the $(A_{1g} + B_{3u})\otimes b_{3u}$ problem of the PJTE in the series of four hexa-heterocycles 1,4-$C_4N_2H_4X_2$ (X = H, F, Cl, Br). (Reprinted with permission from Ref. [85]. Copyright 2015 Elsevier Publishing).

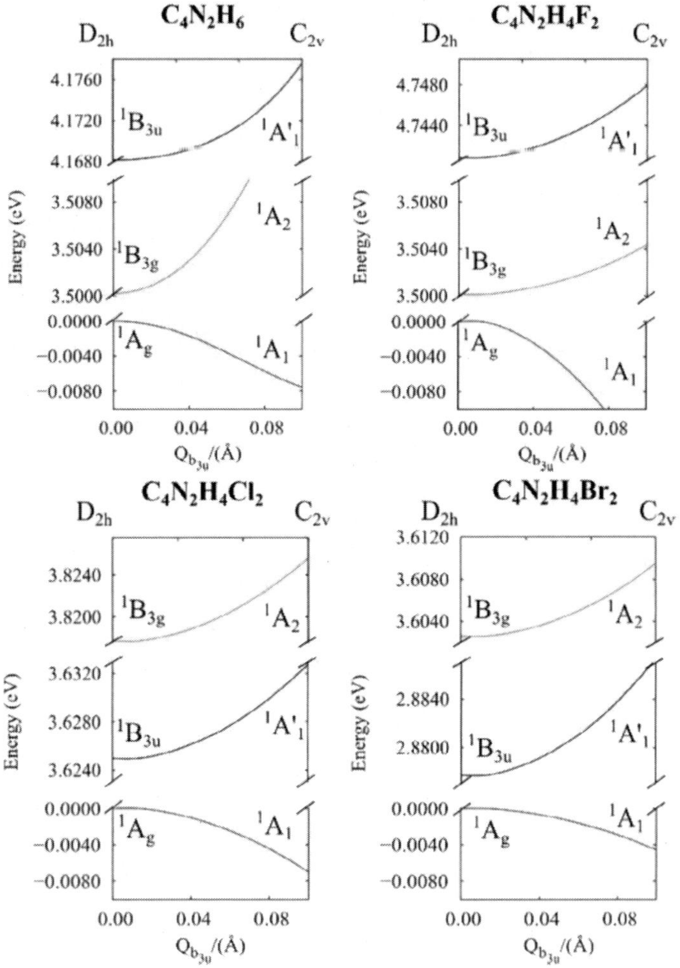

Figure 20. (Color online). Energy profiles (cross sections of the APES) along the coordinate of instability b_{3u} of the planar D_{2h} configuration of the four hexa-heterocycle systems toward the distorted (buckled) C_{2v} structure. (Reprinted with permission from Ref. [85]. Copyright 2014 Elsevier Publishing).

The next step in the evaluation of the PJTE is to reveal the energy gap to the active excited state and orbital excitations of the ground planar configuration that result in this excited state, as well as the parameters of the vibronic coupling between the two mixing states. For this reason ab initio calculations of the energy profiles (cross sections of the APES) of the systems along the instability coordinate b_{3u} were performed. The results are shown in Figure 20. We see that in all the four compounds the unstable along b_{3u} ground state $^1A_{1g}$ is accompanied by a relatively low-lying excited $^1B_{3u}$ state forming a PJTE interacting pair with curvatures assumed to follow a 2 x 2 secular equation. However, there may be higher-in-energy $^1B_{3u}$ states that contribute significantly to the instability of the ground state and, although less probable, it is not excluded that a higher excited is the main reason of instability (sought for in the process of manipulation). To estimate the possible contribution of higher excited states, a negative correction $-p$ to the primary force constants K_0 introduced by the upper states is taken into account (in the form of the second order perturbation correction [86]), resulting in $K_0 - p$ instead of K_0.

We thus separate the whole spectrum of excited states contributions in two parts: (1) the main contribution of the lowest excited states taken into account in a pseudodegenerate problem with a secular equation of the type, and (2) corrections from all the other excited states that lower the primary force constants of the two interacting states. Denoting these corrections to the two states by p_1 and p_2, respectively, we get the following secular equation of the two-level PJTE problem under consideration:

$$\begin{vmatrix} \frac{1}{2}(K_{01} - p_1)Q^2 - \varepsilon & FQ \\ FQ & \frac{1}{2}(K_{02} - p_2)Q^2 + \Delta - \varepsilon \end{vmatrix} = 0 \qquad (32)$$

with the solution,

$$\varepsilon_{1,2} = \frac{1}{4}(K_{01} - p_1 + K_{02} - p_2)Q^2 + \frac{\Delta}{2} \pm \frac{1}{2}\sqrt{\left[\frac{1}{2}(K_{01} - p_1 - K_{02} + p_2)Q^2 - \Delta\right]^2 + 4F^2Q^2} \qquad (33)$$

By fitting these equations to the ab initio calculated energy profiles of the two states $^1A_{1g}$ and $^1B_{3u}$ in Figure 20 we get the parameter values listed in Table 3. Although the separate calculation of the p_1 and p_2 corrections is difficult, they can be extracted from the data in Table 3 by estimating the K_0 values after Eq. (3), using a special program [86]. But for the problem under consideration it is just important that K_{01}-p_1>0, showing that the higher excited states by themselves do not produce the instability of the ground state (the force constant is just softened, but remains positive), meaning the lowest excited state $^1B_{3u}$ is responsible for the main contributor to the instability, and hence the higher states are not the main target for manipulation. The calculations [85] show also the MO populations in the active ground and excited states to be targeted in attempts to suppress the PJTE in these compounds and to flatten their configuration.

Table 3. PJTE coupling constant F (in eV/Å), energy gaps Δ (in eV), and corrected primary force constants $K_1 - p_1$ and $K_2 - p_2$ (in eV/Å^2) of the four molecules obtained by fitting Eq. (20) to the calculated energy profiles in Figure 20 [85]

Molecules	$K_1 - p_1$	$K_2 - p_2$	F	Δ
$C_4N_2H_6$	0.25	0.59	1.42	4.16
$C_4N_2H_2F_2$	0.41	0.38	1.72	4.74
$C_4N_2H_2Cl_2$	0.31	0.56	1.64	3.81
$C_4N_2H_2Br_2$	0.29	0.63	1.56	3.60

7.3. Flattening Puckered (Buckled) 2D and Quasi-2D Systems by Suppressing the PJTE

Following the described above mechanism of the PJTE-induced out-of-plan distortions of the potentially planar configuration of a molecular system, we can see in more detail the possibilities to restore the planar configuration or, more general, to manipulate the planarity of the system by changing its PJTE parameters in the condition of instability (11) by means of targeted external influence. As mentioned above, the relatively easiest way to do this is: (1) increasing the energy gap between the active in the PJTE ground and excited electronic states, e.g., by weak coordination with outside atoms or atomic groups, (2) changing the symmetry

of the active ground or excited states by removing or adding electrons in a redox process, (3) making appropriate chemical substitutions that preserve the 2D structure but restores its planarity, (4) by spectroscopic excitations, (5) applying electric or magnetic fields. Below we review the publications in which some of these possibilities in flattening puckered (buckled) molecular configurations are realized [85, 87-95, 98-103].

7.3.1. Triple-Decker Sandwich Structures CpMoE$_6$MoCp (E = P, As, Sb; Cp = C$_5$H$_5$)

In line with the outlined above methodology of suppressing the PJTE and restoring planar configurations of buckled molecular systems by coordination to other donor or acceptor groups (e.g., via intercalations) is the reported analysis of the triple-decker sandwich structures of a series of complexes [90] $CpMoE_6MoCp$ (E = P, As, Sb; Cp = C_5H_5). In these systems the E_6 groups were shown to occupy the middle-deck position with a planar regular-hexagonal configuration. Meanwhile, these free E_6 groups may hardly be planar. The calculations show that they have several distorted configurations with much lower energy; for P_6 the lowest distorted configuration is by 34.9 kkal lower in energy than the regular hexagonal one.

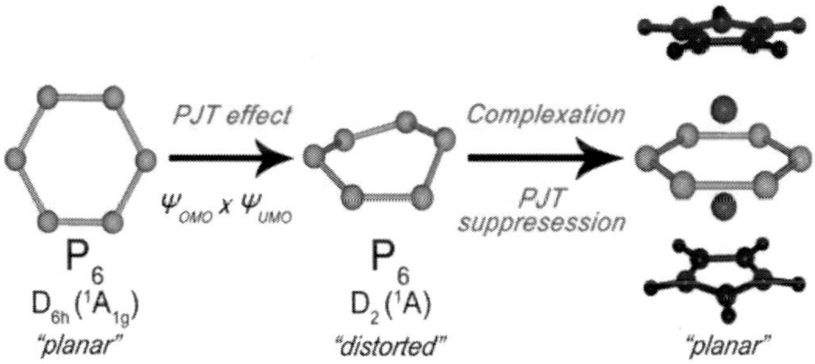

Figure 21. (Color online). Illustration to the suppression of the PJTE in the P_6 ring by its complexation in a triple-dekker system $CpMoE_6MoCp$ (Reprinted with permission from Ref. [90]. Copyright 2012 American Chemical Society).

As expected, the distortion of the latter is due to the PJTE. The calculations [90] shows the interacting HOMO and LUMO that produce the e_{2u} distortion of the P_6 planar hexagonal ring of phosphorous. It follows that in the triple-decker sandwich structure where the E_6 fragments become planar, the PJTE in this fragment is fully suppressed, as illustrated in Figure 21. The detailed ab initio calculations with geometry optimization confirm this conclusion, showing that the active LUMO responsible for the PJTE distortion of the E_6 ring become populated by the molybdenum electrons [90] (interaction between equally populated ground and excited orbitals does not contribute to the PJTE). This is an example when the PJTE is suppressed by coordination to outside species.

7.3.2. Tetra-Heterocyclic 1,2-Diazetes $C_2N_2E_4$, E = H, F, Cl, Br

Similar to deactivating the (active in the PJTE) excited states by populating LUMO with electrons, the condition of PJTE instability (11) can be violated by removing electrons from the ground state. This possibility was checked-out directly by ab initio calculations performed on a series of tetra-heterocyclic 1,2-diazetes [91] $C_2N_2E_4$, E = H, F, Cl, Br (see also [92]). Their square-planar C_{2v} configuration is unstable with respect to two symmetrized displacements a_2 and b_1 due to the two possible PJTE problems, $(^1A_1 + {}^1A_2)\otimes a_2$ and $(^1A_1 + {}^1B_1)\otimes b_1$, the former one along the a_2 puckering displacements leading to the global minimum of the stable configuration C_2 (Figure 22). To reveal external redox influence on this PJTE instability and distortions, ab initio calculations of the four tetra-heterocyclic 1,2-diazetes, $C_2N_2E_4$, E = H, F, Cl, Br, in their cation, anion, dication, dianion, and triplet (HOMO–LUMO) states of the planar configuration were performed, and their a_2- and b_1-type PJTE instabilities were estimated by the corresponding imaginary frequencies in the high-symmetry C_{2v} configuration [91]. The results are shown in Table 4.

We see that in the cation states the instability decreases, and three of the four systems become stable in the dication state. $C_2N_2Br_4$ is unstable also in the excited triplet state. The first (more obvious) reason for the weakening of the PJTE in these cases is the much larger energy gap Δ between the (changed) ground $^1A_1$ and active excited $^1A_2$ states. The calculated new energy gaps Δ in the dications $C_2N_2E_4^{2+}$ are (in eV): 10.81, 9.65, 7.89, and 5.72 for E = H, F, Cl, and Br, respectively, compared with 7.03, 5.67, 5.09, and 4.08 for the neutral molecules.

The other parameters in the condition of instability (12) include the primary force constant K_0 which increases by removing electrons (inter-electron repulsions *ceteris paribus* soften the system), thus weakening the

PJTE too, while the vibronic coupling constant F formed by similar electronic states is expected to remain of the same order of magnitude.

Table 4. Imaginary frequencies (in cm^{-1}) along a_2 and b_1 normal displacements in the cation, anion, dication, dianion, and excited triplet states of the $C_2N_2E_4$ series (dashes indicate no-imaginary frequencies). In the neutral (unstable) C_{2v} configuration, the imaginary a_2 frequencies of the four compounds are 882.5, 838.3, 511.9, and 481.2, respectively [91]

Molecule	**Cations** $C_2N_2E_4^+$	**Anions** $C_2N_2E_4^-$	**Triplet state HOMO→LUMO**	**Dications** $C_2N_2E_4^{2+}$	**Dianions** $C_2N_2E_4^{2-}$
Imag. Freq. (cm^{-1})	a_2/b_1	a_2/b_1	a_2/b_1	a_2/b_1	a_2/b_1
$C_2N_2H_4$	454.2/ --	999.8/939.3	1069.3/978.7	-- / --	942.1/971.6
$C_2N_2F_4$	600.1/489.4	1079.0/839.1	685.8/119.5	117.8/ --	Problem in calculation
$C_2N_2Cl_4$	327.9/213.8	419.3/351.2	873.4/689.7	-- / --	435.2/409.0
$C_2N_2Br_4$	293.8/189.7	665.6/53.75	-- / --	-- / --	45.0/121.3

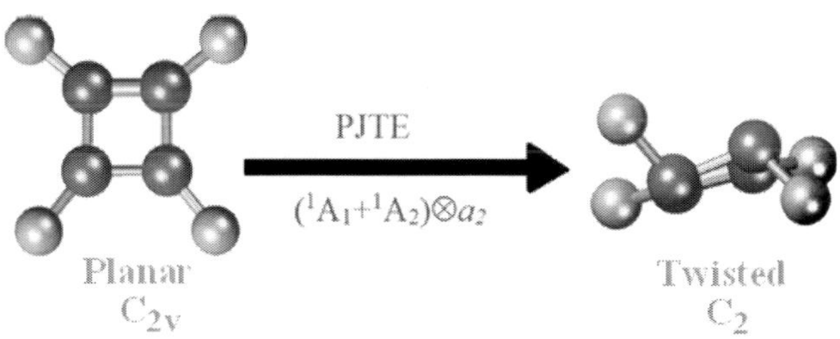

Figure 22. (Color online). Illustration to the $C_{2v} \rightarrow C_2$ puckering transformations of the $C_2N_2E_4$ systems resulting from the $(^1A_1 + ^1A_2)\otimes a_2$ PJTE problem (Reprinted with permission from Ref. [91]. Copyright 2015 Elsevier Publishing).

On the basis of these calculations the authors [91] suggested that the planar configurations of the $C_2N_2E_4$ (E = H, F, Cl, Br) heterocycle compounds, distorted by the PJTE, can be restored in the dication state, produced, for instance, by an inverse sandwich structure with two additional electron acceptors on both sides of the neutral cycle, similar to some suggestions for other systems, shown above [89, 90].

7.3.3. 2D-Silicene

Silicene is the nearest analog of graphene, and therefore it got elevated attention in search for useful 2D materials. Similar to graphene, silicene has a hexa-ring structure, but distinguished from graphene, the silicene rings are not planar, but puckered, with a puckered angle of 38^0 (Figure 23). As expected based on the general theory, the puckering of the planar configuration of silicene was shown to result from the PJTE (Figure 24), the latter being the basic reason of its structural difference from graphene [93, 94]. Figure 25 illustrates the results obtained for the analog of the silicene local structure, hexasilabenzene Si_6H_6, which has the same hexagonal Si_6 ring.

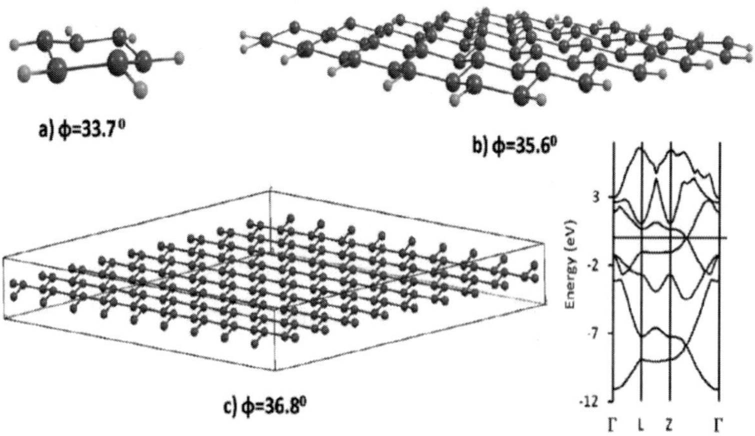

Figure 23. Puckering angle Φ for (a) Si_6H_6, (b) $Si_{70}H_{22}$, and (c) silicene sheet and its band structure (Reprinted with permission from Ref. [93]. Copyright 2012 American Chemical Society).

The PJT vibronic coupling between its HOMO e_{1g} and LUMO e_{2g} results in the b_{2g} type distortion transforming the D_{6h} structure into a D_{3d} one (Figure 25*a*). Then they show that two positive ions Li+ arranged in an inverse sandwich geometry fully suppress the PJTE distortion (Figure 25*b*). The calculations [93] show also that the interaction of the Li+ ions with the silicon ring is mostly ionic, and the PJTE is suppressed because of the increase of the HOMO-LUMO energy gap Δ between the PJT interacting states from 6.83 eV to 7.68 eV, which destroys the inequality (11).

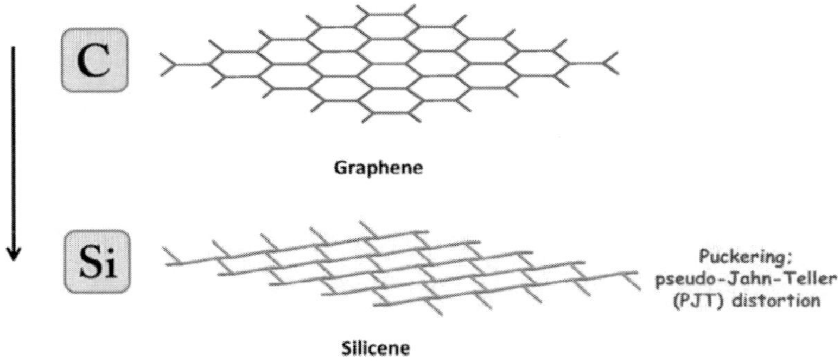

Figure 24. Illustration to the difference in the structures of graphene and silicene induced by the PJTE in the latter (in graphene there is no PJTE instability of the planar configuration). (Reprinted with permission from Ref. [94]. Copyright 2014 American Chemical Society).

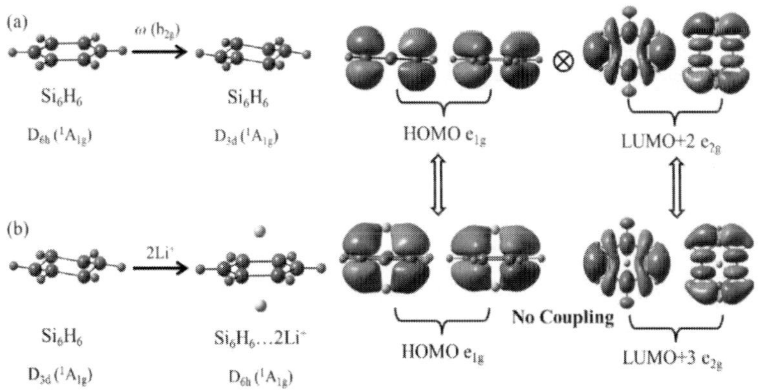

Figure 25. (Color online). (a) Onset of the puckering distortion $D_{6h} \rightarrow D_{3d}$ via b_{2g} displacements in hexasilabenzene, and the HOMO and LUMO involved in the PJTE; (b) Suppression of the ring distortion in the hexasilabenzene-$2Li^+$ complex: with the HOMO and LUMO of same symmetry the PJTE is suppressed (shown as "no coupling") because of the significant increase of the energy gap between them (Reprinted with permission from Ref. 94. Copyright 2014 American Chemical Society).

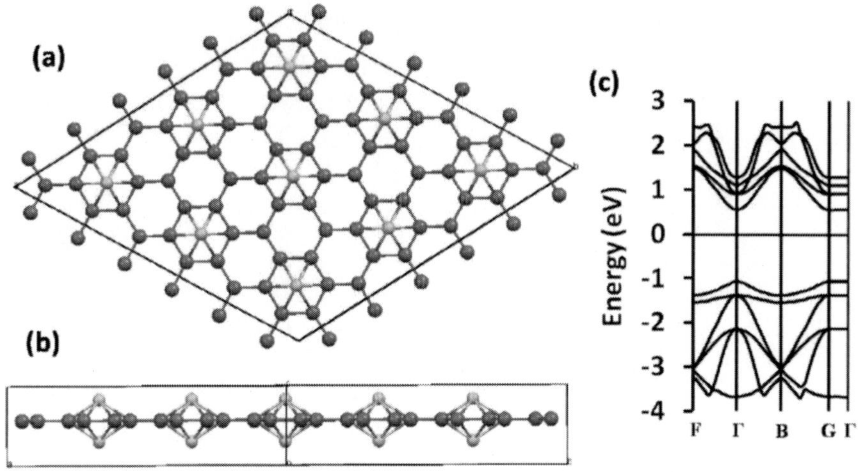

Figure 26. (Color online).Optimized structure of silicene with appropriately intercalated Li^+ ions: (a) transverse view, (b) lateral view, (c) electronic band structure (Reprinted with permission from Ref. [94]. Copyright 2014 American Chemical Society).

The authors [93, 94] discuss also the implications that occur in the extended systems and suggest a virtual (an ab initio optimized) structure of silicene with intercalated Li^+ ions (Figure 26), in which the PJTE is suppressed and the Si atoms form regular hexagons in a planar arrangement. Mixed carbon-silicon ring systems are also considered in this paper.

7.3.4. 2D-Phosphorene

Similar to silicene, phosphorene (Pn) can be viewed as a layered quasi 2D system composed by buckled hexagonal rings, but unlike silicene, the heavier P atoms in Pn produce smaller energy gaps to excited states resulting in several stable conformations of lower symmetry [95, 96]. Among the latter the two stable buckled configuration known as black and blue phosphorene are most important in view of the 2D systems under consideration. As above, the buckling of Pn can be conveniently studied using the P_6H_6 molecule as a model unit [95]. The planar regular-hexagonal D_{6h} configuration of P_6H_6 is unstable due to the PJTE; as shown by calculations, there are two most active PJT displacements, e_{1g} and b_{2g}, leading to two stable distorted configurations of C_{2h} and D_{3d} symmetry. It is notable that the basic P_6 skeletons in these two molecular configurations of P_6H_6 coincide well with the ones in black and blue phosphorene, respectively (Figure 27). Hence the PJTE origin of the buckling in the molecular system can be employed to explain the origin of the buckling in the extended systems of black and blue Pn and to manipulate their properties. The authors [95] suggested to use several molecular systems (shown in Figure 28) as molecular dopants in order to influence the structural properties of the two phosphorenes by targeting their PJTE parameters of the latter with the known donor-acceptor properties of the former (molecules 1–4 are known to be donors, while 5 and 6 are acceptors).

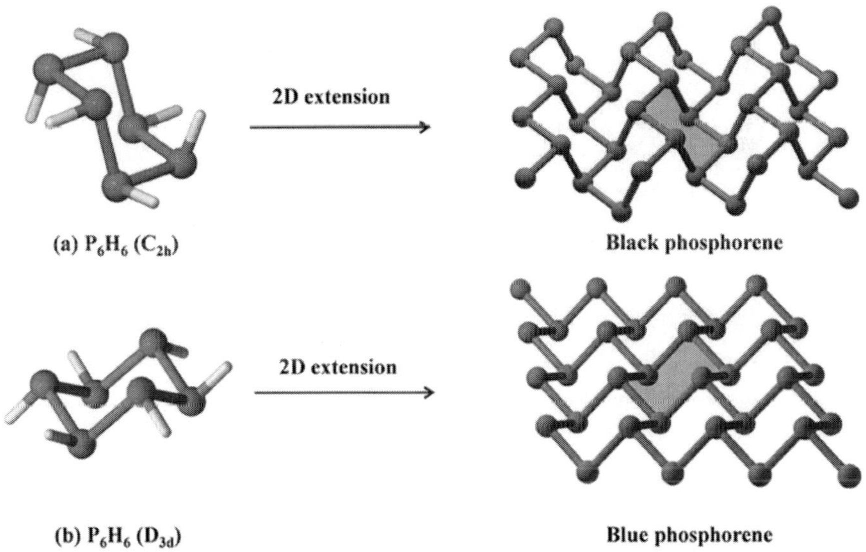

Figure 27. (Color online). One-to-one mapping between the two PJTE-induced molecular distortions in P_6H_6, e_{1g} and b_{2g}, and the corresponding two buckling forms of phosphorene (Pn), black and blue, respectively: (a) the e_{1g} type distorted configuration of the P_6 ring in P_6H_6 with C_{2h} symmetry is the same as in the local P_6 rings in black Pn; (b) the D_{3d} configuration of the P_6 ring in P_6H_6 due to its PJTE b_{2g} type distortion from the planar regular-hexagonal geometry is the same as the local P_6 ring configuration of blue Pn. The distortion parameters in the molecular and extended systems are very close, in both (a) and (b) cases (the shaded regions in the pictures are shown to ease the visual structural comparisons) (Reprinted with permission from Ref. [95]. Copyright 2016 American Chemical Society).

Table 5. Band gaps and charge transfer Δq (in electron charge (|e|) for pristine black and pristine blue phosphorenes (Pn) and their respective complexes with dopant molecules (numbered in Figure 28) in the most stable sites of adsorption. Positive and negative values of Δq indicate the amount of electronic charge added to and removed from the dopant, respectively [95]

System	Molecule	Band gape (eV)	Charge transfer Δq (|e|)
Black Pn	-	0.87	-
	4	0.19	-0.23
Blue Pn	-	1.98	-
	3	0.27	-0.17
	4	0.08	-0.78
	5	0.70	+0.50
	6	0.64	+0.61

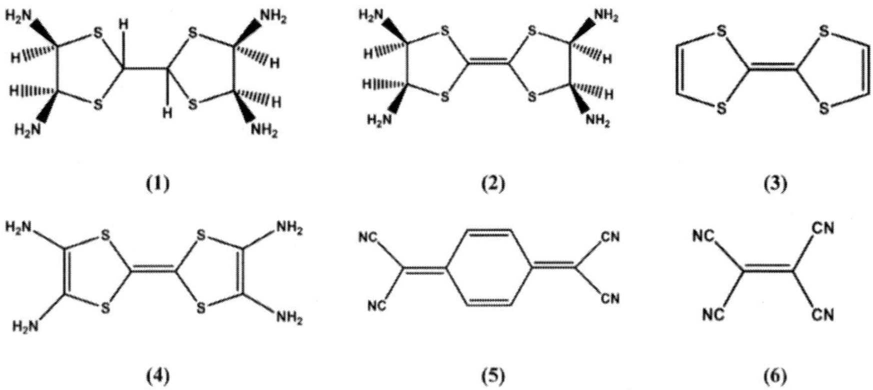

Figure 28. Molecules used as phosphorene dopants: (1) 6H-TTF (hexahydro-tetratiofulvalene), (2) 2H-TTF (dihydro-TTF), (3) TTF, (4) 4NH2-TTF (tetraamino-TTF), (5) TCNQ (tetracyanoquinodimethane), and (6) TCNE (tetracyanoethylene).

To explore this possibility, calculations of charge distribution and charge transfers by coordination (adsorption) of these molecules to the surface of black and blue phosphorene, as well as their influence on the electronic band structure and band gaps, were carried out [95] (see also [97]). A part of the results are illustrated in Figure 29 with numerical estimates in Table 5. It emerges from these data and electronic band structures that the dopant molecules indeed influence significantly the properties of the Pn systems by either donating (molecules 3 and 4), or accepting (molecules 5 and 6) electronic charge, thus modifying the PJT interacting ground-excited states. It can be seen that, first, blue Pn is better exposed for interacting with the dopants then black Pn. This is due to the smaller buckling amplitude in the blue Pn and hence better contact interaction (orbital overlaps) between dopant and Pn atoms. We see that the strong donor molecule tetraamino-tetratiofulvalene strongly reduces the energy gap value in both black and blue Pn, thus transforming them into n-semiconductors. The detailed analysis of the electronic band structure [95] shows how the dopants modify the highest occupied and lowest unoccupied energy levels that form the band structure with corresponding changes in observable properties.

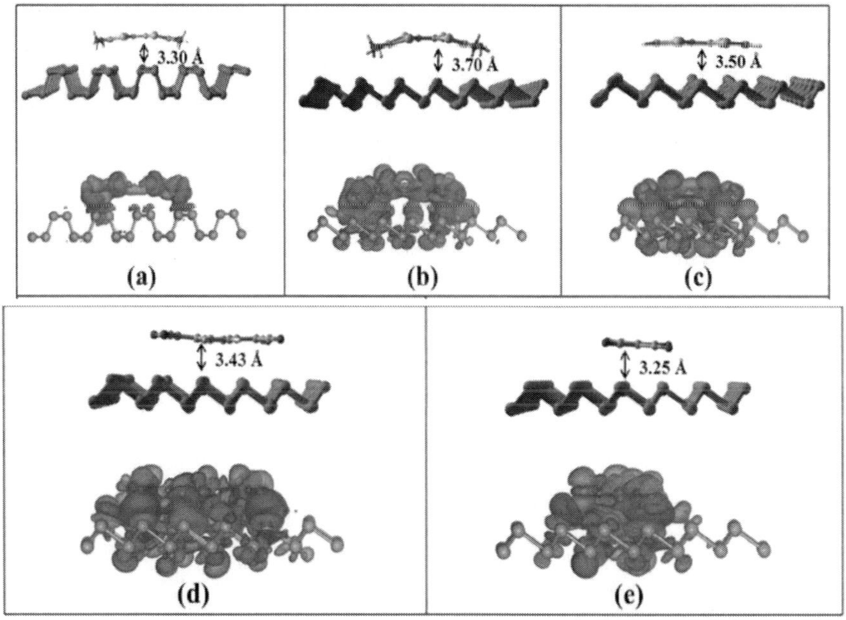

Figure 29. (Color online). Charge density differences (CDD) in phosphorene Pn plus the adsorbed dopant molecule numbered in Figure 28: (a) black Pn + (4), (b) blue Pn + (4), (c) blue Pn + (3), (d) blue Pn + (5), and (e) blue Pn + (6), plotted with isosurface value of 0.026 |e|/Bohr^3. The top panel for each entry shows the average molecule-surface distances in the most stable configurations. CDD is calculated by subtracting the charge density of fragments from the total charge density of the system. Positive and negative CDD are indicated by cyan and olive colors, respectively (Reprinted with permission from Ref. [95]. Copyright 2016 American Chemical Society).

7.3.5. 2D-Germanene

Similar to silicene, germanene consists of buckled hexa-rings with the buckling amplitude significantly bigger than in silicene; the maximal buckling distance in the $D_{6h} \rightarrow D_{3d}$ distortion of the hexa-ring in Ge_6H_6 is d = 0.61 Å. As in all the other cases of spontaneous distortion of high-symmetry configurations, this buckling is shown to be due to the PJTE induced by the vibronic coupling of the electronic ground state A_{1g} with the low-lying excited state B_{2g} under the nuclear displacements of b_{2g} symmetry [98]. Distinguished from the silicene case the authors suggest another way to influence the germanene properties, namely by hydrogenation to form germanane, in which a large energy gap is opening (Figure 30). This opens a way to control the properties of germanene by further chemical substitutions.

7.3.6. Dithiine, Thiantrene, and Anthracene

The PJTE origin of distortions from the planar configuration (puckering) and some of the mentioned above methods of restoration of the planar configuration were demonstrated also on two sets of 1,2- and 1,4-dithiine containing tricyclic compounds taken as examples [99, 100]. In the first set the neutral carbon sulfide C_6S_8 and its dianion $C_6S_8^{2-}$, containing two S-atoms in the 1,2-positions of the central six-membered ring and one thione (C = S) bond in the five-membered rings on its either side were considered. Both compounds and their structure were studied experimentally and confirmed by single-crystal X-ray diffraction [101, 102].

Crystallographic data show that the neutral C_6S_8 compound is not planar, it has a puckered structure of C_2 symmetry with a twisted 1,2-dithiin core, and its reduction has resulted in cleavage of the SAS bond and rotation of the two peripheral rings about the central C-C bond [101, 102]. The authors [99, 100] show that in both systems the out-of-plane distortions are due to the PJT coupling between the ground A_1 and two excited electronic states of A_2 symmetry, and that by oxidation the PJTE in C_6S_8 is weakened, being completely suppressed in $C_6S_8^{2+}$. The 1,2-dithiine compound is notable also in what it was shown as an example of another mechanism of suppressing the PJTE in 2D systems by aggregation due to intermolecular p-stacking interactions of planar configuration [99].

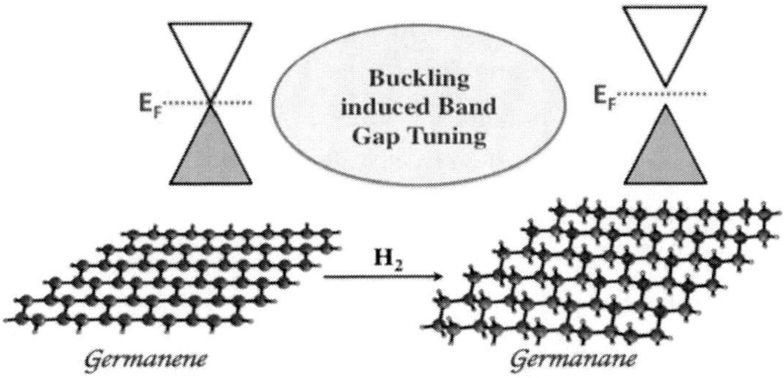

Figure 30. (Color online).Transformation of germanene into germanane by hydrogenations opens new ways to control the energy gap and other properties of the 2D system by means of consequent chemical substitutions (Reprinted with permission from Ref. [98]. Copyright 2015 American Chemical Society).

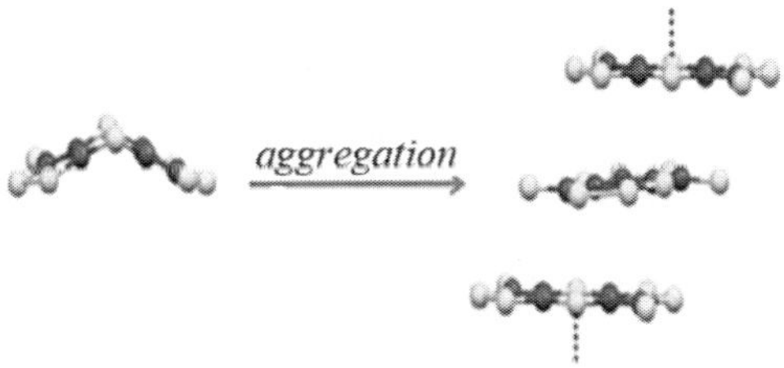

Figure 31. (Color online). The free molecule of 1,2-dithiine C_6S_8 (left) has a puckered structure of C_2 symmetry (induced by the PJTE in the planar high-symmetry configuration C_{2h}), which is fully suppressed in the crystal structure of the α polymorph due to the π-stacking interaction between at least three molecules in a planar-parallel arrangement (right) (Adapted with permission from Ref. [99]. Copyright 2015 American Chemical Society).

Figure 31 illustrates this mechanism. Performing detailed ab initio calculations the authors show that the p-staking interactions between parallel planar configuration of the C_6S_8 molecule contributes to the weakening of the PJT distortions in each of them, making them less distorted, and reaching an exact planar configuration in a trimer parallel arrangement shown in Figure 31. This explains the origin of the $C_2 \rightarrow C_{2h}$ transformation of the molecular geometry when moving from the free molecule to the α polymorph of the crystal state. The C_6S_8 compound in the solid state has also another polymorph, β, for which the distorted configuration of the free molecule remains distorted in the crystal structure.

A similar phenomenon of flattening by aggregation takes place in a more extended quasi planar compound dithianon, $C_{14}H_4N_2O_2S_2$, which is distorted in the free molecule, but becomes flattened by aggregation in one of its polymorphs. In this case, as shown by calculations [99], the parallel stacking of only two molecule is sufficient to destroy the PJTE induced out-off-plan distortion.

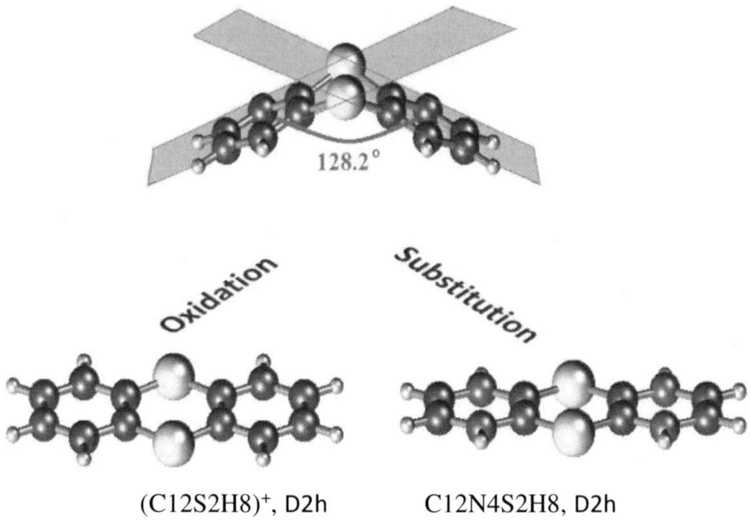

Figure 32. (Color online). Thianthrene is non-planar in the ground state due to the PJTE. Removing an outer electron quenches the latter and restores the planar configuration. Alternatively, substitution of four carbon atoms by nitrogen produces a similar effect of planarization (Reprinted with permission from Ref. [100]. Copyright 2017, Journal of Physics).

The second set of studied tricyclic systems contains thianthrene, or dibenzo-1,4-dithiin molecule, and anthracene, and their derivatives [100]. According to crystal structure determinations the thianthrene molecule is bent at the S-S axis by 127– 128, and it has a low energy flapping vibration about the S-S axis, while the oxidized thianthrene has a planar nuclear configuration with D_{2h} symmetry. Besides, if the C atoms at 1, 4, 6, and 9 positions are substituted by nitrogen atoms, the dithiin core becomes planar (Figure 32). On the other hand, the anthracene molecule (An) is planar with the nuclear configuration of D_{2h} symmetry, whereas the 9,10-dihydroanthracene (9,10-H_2An) has a folding geometry of C_{2v} symmetry with a dihedral angle of about 150–160_. These structural features are fully controlled by the PJTE, its mechanism being demonstrated in detail by ab initio calculation [100]. For all the considered molecules the geometry optimization and vibrational frequencies in the unstable planar and equilibrium nuclear configurations were carried out, and their ground and low-lying excited energy terms, the HOMO and LUMO and their changes along the distortion coordinates were analyzed to show how the observed PJTE distortions emerge and how they can be manipulated. The main instruments used to do so are oxidation, reduction, and atomic substitution. Figure 32 illustrates the two ways to flatten the thiantrene molecule by suppressing the PJTE via oxidation or substitution [100].

7.3.7. Triazine-Based 2D-Graphitic Carbon Nitride (g-C_3N_4)

Graphitic carbon nitride g-C3N4 is a 2D system which is similar to graphene in many aspects, but unlike graphene g-C3N4 is not regular planar, it is buckled as shown in Figure 33 for a whole sheet, and in Figure 34 in a cluster unit presentation [103]. The PJTE origin of the buckling in the cluster unit $C_9N_{12}H_6$ (Figure 34) was studied in detail with the goal to influence it by external perturbations [103]. It was shown that its planar configuration is unstable along the *e″* symmetrized displacements for which the coupled HOMO and LUMO responsible for the instability were revealed by ab initio calculations. With these data and previous experience, the authors [103] suggested suppressing the PJTE and restoring the planar configuration of this system by adding Be^{2+} ions below and above the fragment unit in an inverse sandwich arrangement, similar to the considered above cyclohexasilane case. The calculations show indeed that the electron acceptor ions change the HOMO-LUMO positions, thus quenching the PJTE and flattening the system. Figure 35 shows the optimized structures of both the fragment unit and the extended sheet of g-C_3N_4 with appropriately positions of the Be^{2+} ions which restore the planar configuration, showing also the electronic band structure.

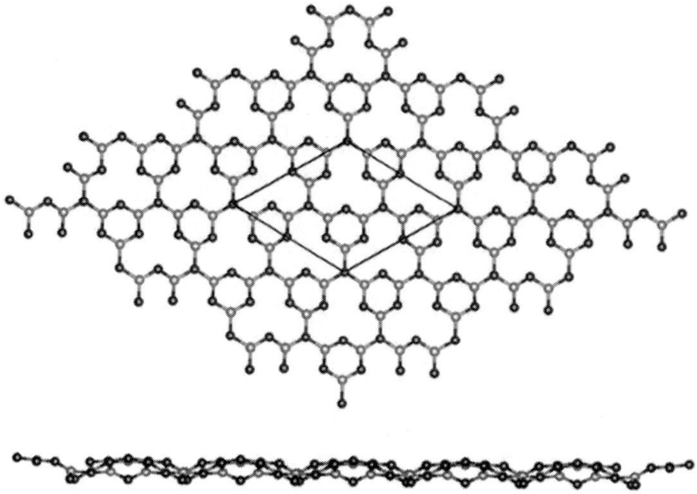

Figure 33. (Color online). Buckled 2D structure of graphitic carbon nitride *g*-C_3N_4 shown in top and side view. Here and below in Figure 34 the gray, blue, and white spheres represent carbon, nitrogen, and hydrogen atoms, respectively (Reprinted with permission from Ref. [103]. Copyright 2015 American Chemical Society).

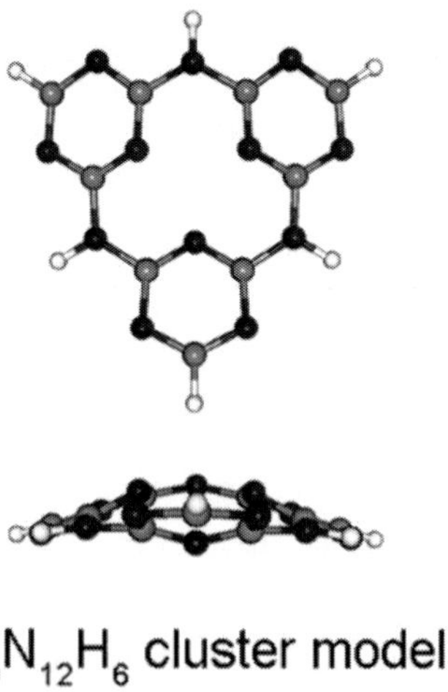

Figure 34. (Color online). Local buckling unit of graphitic carbon nitride *g*-C_3N_4 that is repeated along the extended 2D system in Figure 33 (Reprinted with permission from Ref. [103]. Copyright 2015 American Chemical Society).

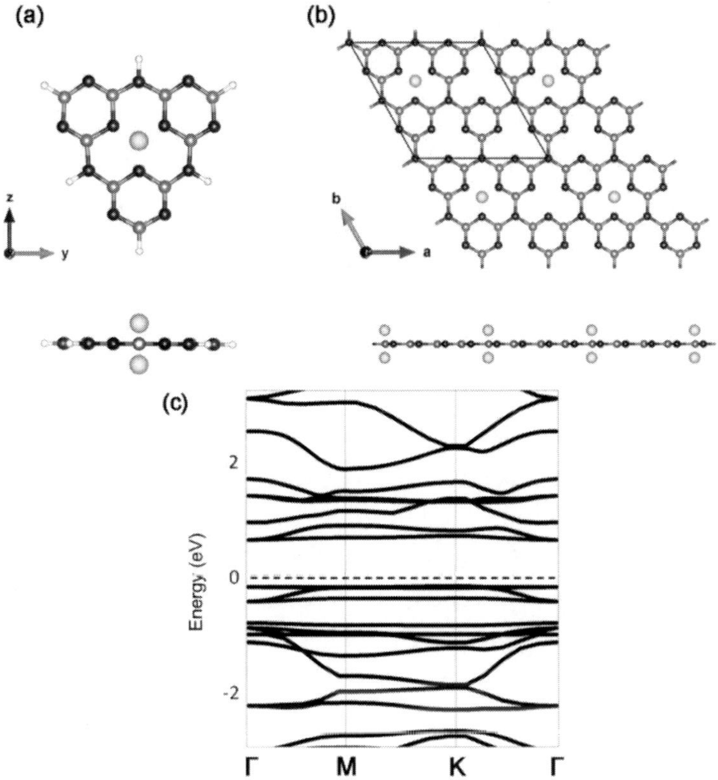

Figure 35. (Color online). Top and side views of the optimized complex of the graphitic carbon nitride g-C_3N_4 local unit with two Be^{2+} ions in an inverse sandwich arrangement (a), the corresponding extended 2D sheet (b) and its electronic band structure (c) [103] (Reprinted with permission from Ref. 103. Copyright 2015 American Chemical Society).

CONCLUSION

The chapter introduces a new trend in search of enhanced or new materials with properties important for applications, by exploiting one of the fundamental features of matter known as the JTE or PJTE. Based on the special local distortion dynamics, induced by these effects, several lines of applications are suggested, some of which were already shown to be most efficient. Even more important is that this trend opens a way of unlimited possibilities for further developments along an almost unlimited variety of possible materials with useful properties; there are no *a priory* limitations in materials that may have the JTE and especially the PJTE.

References

[1] Bersuker, I. B., *The Jahn-Teller Effect*, Cambridge University Press, Cambridge UK, 2006.

[2] Englman, R., *The Jahn-Teller Effect in Molecules and Crystals*, Wiley, New York, 1972.

[3] Bersuker, I. B., The Pseudo Jahn-Teller Effect – a Two-State Paradigm in Formation, Deformation, and Transformation of Molecular Systems and Solids, *Chem. Rev*. 113, 1351 (2013).

[4] Bersuker, I. B., and Polinger, V. Z., *Vibronic interactions in Molecules and Crystals,* Springer-Verlag, Berlin-Heidelberg, 1989.

[5] Bersuker, I. B., Spontaneous Symmetry Breaking in Matter Induced by Degeneracy and Pseudodegeneracy, in *Advances in Chemical Physics*, Vol. 160, S. Rice and R. Dinner, eds., Wiley, New York, 2016, pp 159-208.

[6] Bersuker, I. B., *J. Phys. Conf. Series*, 833, 012001 (2017); *Flat Chem,* 6, 11–27 (2017).

[7] Bersuker, I. B., *Phys. Lett.* 20, 589 (1966).

[8] Polinger, V. Z., Garcia-Fernandez, P., and Bersuker, I. B., *Physica B. Condens. Mater*, 457, 296 (2015).

[9] Bersuker, I. B., The vibronic (pseudo Jahn-Teller) theory of ferroelectricity. Novel aspects and applications, *Ferroelectrics*, 536, 1-59 (2018).

[10] Polinger, V. Z., *J. Phys. Conf. Series* 428, 012026 (2013).

[11] Comes, R., Lambert, M., and Guinner, A., *Solid State Commun.* 6, 715 (1968); 12, 1053 (1973); Comes, R., Currat, R., Denoyer, F., Lambert, M., and Quittet, M., *Ferroelectrics* 12, 3 (1976).

[12] A letter states: "University of Paris…Orsay, January 23 [1969], …Dear Dr Bersuker, We just discovered your note "On the Origin of Ferroelectricity in Perovskite type Crystals" published in *Physics*

Letters (1. April 1966). From a completely different approach we came to conclusions, which are similar to yours. You will find enclosed our publications on $BaTiO_3$ and $KNbO_3$… Sincerely Yours, R. Comes."

[13] Blinc, R., Zalar, B., Laguta, V. V., and Itoh, M., *Phys. Rev. Lett.* 94, 147601 (2005).

[14] Ravel, B., Stern, E. A., Vedrinskii, R. I., and Kraizman, V., *Ferroelectrics* 206-207, 407 (1998); E. Stern, *Phys. Rev. Lett.* 93, 037601 (2004).

[15] Müller, K. A., and Berlinger, W., *Phys. Rev. B* 34, 6130 (1986); Müller, K. A., Berlinger, W., Blazey, K. W., and Albers, J., *Solid State Commun.* 61, 21 (1987); Volkel G., and Müller, K. A., *Phys. Rev. B* 76, 094105 (2007).

[16] Harada, J., Axe, J. D., and Shirane, G., *Phys. Rev. B* 4, 155 (1971); Luspin, Y., Servoin, J. L., and Gervais, F., *J. Phys. C* 13, 3761 (1980); Vogt, H., Sanjurjo J. A., and Rossbroich, G., *Phys. Rev. B* 26, 5904 (1982).

[17] Barone, Paolo, and Picozzi, Silvia*, C. R. Physique* 16, 143 (2015).

[18] Bersuker, I. B., *Phys. Rev. Lett.,* 108, 137202 (2012).

[19] Bersuker, I. B., *Electronic Structure and Properties of Transition Metal Compounds. Introduction to the Theory*, second ed., Wiley, New York, 2010.

[20] Gütlich, P., and Goodwin, H. A., eds., *Spin Crossover in Transition Metal Compounds*, Springer-Verlag, Berlin, Heidelberg, 2004, vol. I, II, and III.

[21] Hauser, A., Enachescu, M. L., Daku, D., Vargas, A., Amstutz, N., Low-temperature lifetimes of metastable high-spin states in spin-crossover and in low-spin iron(II) compounds: The rule and exceptions to the rule, *Coord. Chem. Rev.* 250, 1642-1652 (2006).

[22] Domracheva, N. E., Pyataev, A. V., Vorobeva, V. E., and Zueva, E. M., *Phys. Chem. B* 117, 7833 (2013).

[23] Raymond, O., Ostos, C., Font, R., Curiel, M., Bueno-Baques, D., Machorro, R., Mestres, L., Portelles, J., and Siqueiros, J. M., *Acta Mater.* 66, 184 (2014).

[24] Weston, L., Cui, X. Y., Ringer, S. P., and Stampfl, C., *Phys. Rev. B*, 93, 165210 (2016).

[25] Bokov, A. A., and Ye, Z.-G., *J. Mater. Sci.* 41, 31 (2006); *J. Adv. Dielectr.* 02, 1241010 (2012); Wang, D., Bokov, A. A., Ye, Z.-G., Hlinka, J., and Bellaiche, L., *Nature Commun.,* 7, 11014 (2016).

[26] Cowley, R. A., Gvasaliya, S. N., Lushnikov, S. G., Roessli, B., and Rotaru, G. M., *Adv. Phys.* 60, 229 (2011).

[27] Samara, G. A., in *Solid State Physics*: *Advances in Research and Applications*, eds. Ehrenreich, H., and Spaepen F., Academic, New York, 2001, Vol. 56, p. 1; *J. Phys.: Condens. Matter* 15, R367 (2003).

[28] Xu, G., Shirane, G., Copley, R. D., and Gehring, P. M., *Phys. Rev. B* 69, 064112 (2004).

[29] Burns, G., and Dacol, F. H., *Solid State Commun.* 48, 853 (1983).

[30] Fisch, R., *Phys. Rev. B* 67, 094110 (2003).

[31] Colla, E. V., Griffin, P., Delgado, M., Weissman, M. B., Long, X., and Ye, Z.-G., *Phys. Rev. B* 78, 054103 (2008); Delgado, M., Colla, E. V., Griffin, P., Weissman, M. B., Viehland, D., *Phys. Rev. B* 79, 140102(R) (2009).

[32] Kleemann, W., Dec, J., and Miga, S., *Ferroelectrics* 499, 72 (2016).

[33] Vugmeister, B. E., and Glinchuk, M. D., *Rev. Mod. Phys.* 62, 993(1990); Vugmeister, B. E., *Phys. Rev. B* 73, 174117 (2006).

[34] Glinchuk, M. D., Ragulya, A. V., and Stephanovich, V. A., *Nanoferroics*, in *Springer Series in Material Science*, Vol. 177 (Springer Science + Business Media, Dordrecht, 2013), Chap. 3.

[35] Geneste, G., *Phys. Rev. B* 79, 064101 (2009); 79, 144104 (2009).

[36] Takenaka, H., Grinberg, I., Liu, S., and Rappe, A. M., *Nature* 546, 391 (2017).

[37] Polinger, V., and Bersuker, I. B., *Phys. Rev. B*, 98, 214102 (2018).

[38] Bersuker, I. B., *Appl. Phys. Lett.* 106, 022903 (2015).

[39] Bersuker, I. B., *Appl. Phys. Lett.* 107, 202904 (2015).

[40] Polinger, V. Z., and Bersuker, I. B., *J. Phys. Conf. Series,* 833, 012012 (2017).

[41] Kogan, S. M., *Sov. Phys. Solid State* 5, 2069 (1964).

[42] Zubko, P., Catalan, G., and Tagantsev, A. K., Flexoelectric effect in solids, *Annu. Rev. Mater. Res.* 43, 387 (2013).

[43] Yudin, P. V., and Tagantsev A. K., Fundamentals of flexoelectricity in solids, *Nanotechnology* 24 (2013) 432001

[44] Tagantsev A. K., and Yudin, P. V., *Flexoelectricity in Solids. From Theory to Applications,* World Scientific, Singapore, 2017; Morozovska, A. N., Khist, V. V., Glinchuk, M. D., Scherbakov, C. M., Silibin, M. V., Karpinsky, D. V., and Eliseev, E. A., *J. Mol. Liquids* 267, 550 (2018).

[45] Yudin, P. V., Ahluwalia, R., and Tagantsev, A. K., *Appl. Phys. Lett.* 104, 082913 (2014).

[46] Ma, W., *Phys. Status Solid. B* 247, 213 (2010); Ma, W., and Cross, L. E., *Appl. Phys. Lett.* 79, 4420 (2001); 81, 3440 (2002); 88, 232902 (2006).

[47] Nguyen, T. H., Mao, S., Yeh, Y.-W., Purohit, P. K., and McAlpine, M. C., *Adv. Mater.* 25, 946 (2013).

[48] Hong, J., Catalan, G., Scott, J. F., and Artach, E., *J. Phys. Condens. Matter* 22, 112201 (2010).

[49] Biancoli, A., Fancher, C. M., Jones, J. L., and Damjanovic, D., *Nat. Mater.* 14, 224 (2015).

[50] Chen, F., Zhu, Y., Liu, S., Qi, Y., Hwang, H. Y., Brandt, N. C., Lu, J., Quirin, F., Enquist, H., Zalden, P., Hu, T., Goodfellow, J., Sher, M.-J., Hoffmann, M. C., Zhu, D., Lemke, H., Glownia, J., Chollet, M., Damodaran, A. R., Park, J., Cai, Z., Jung, I. W., Highland, M. J., Walko, D. A., Freeland, J. W., Evans, P. G., Vailionis, A., Larsson, J., Nelson, K. A., Rappe, A. M., Sokolowski-Tinten, K., Martin, L. W., Wen, H., and Lindenberg, A. M., *Phys. Rev. B* 94, 180104(R) (2016).

[51] Narvaez, J., and Catalan, G., *Appl. Phys. Lett.* 104, 162903 (2014).

[52] Li, Y., Shu, L., Zhou, Y., Guo, J., Xiang, F., He, L., and Wang, H., *Appl. Phys. Lett.* 103, 142909 (2013).

[53] Burcsu, E., Ravichandran, G., and Bhattacharya, K., *Appl. Phys. Lett.* 77, 1698 (2000).

[54] Surowiak, Z., Fesenko, E. G., and Skulski, R., *Arch. Acoust.* 24, 391 (1999).

[55] Kahn, O., and Martinez, C. J., Spin-transition polymers: from molecular materials toward memory devices, *Science* 279, 44-48 (1998).

[56] Feringa, B. L., van Delden, R. A., Koumura, N., and Geertsema, E. M., Chiroptical molecular switches. *Chem. Rev.* 100, 1789 (2000).

[57] Rocha, A. R. et al., Towards molecular spintronics, *Nat. Mater.* 4, 335 (2005).

[58] Renz, F. et al., Strong Field Iron(II) complex converted by light into a long-lived high-spin state, *Angew. Chem. Int. Ed.* 39, 3699 (2000).

[59] Gatteschi, D., Sessoli, R., and Villain, J., *Molecular Nanomagnets,* Oxford University Press, New York, 2006.

[60] Larionova, J., Salmon, L., Guari, Y., Tokarev, A., Molvinger, K., Molnar, G., and Bousseksouet, A., *Angew. Chem., Int. Ed.* 47, 8236 (2008).

[61] Boldog, I., Gaspar, A. B., Martinez, V., Pardo-Iban˜ez, V., Ksenofontov, V., Bhattacharjee, A., Gutlich, P., and Real, J. A., *Angew. Chem., Int. Ed.* 120, 6533 (2008).

[62] Garcia-Fernandez, P., Bersuker, I. B., and Boggs, J. E., Orbital disproportionation and spin crossover as a pseudo Jahn-Teller effect, *J. Chem. Phys.* 125, 104102 (2006).

[63] Chibotaru, L. F., and Ceulemans, A., *Phys. Rev. B* 53, 15522 (1996); Ceulemans, A., Chibotaru, L. F., and Cimpoesu, F., *Phys. Rev. Lett.* 78, 3725 (1997).

[64] Bersuker, I. B., Recent Developments in the Jahn-Teller Effect theory. The Hidden Jahn-Teller Effect, In *The Jahn-Teller Effect: Fundamentals and Implications for Physics and Chemistry*, Springer Series in Chemical Physics, v. 97, Eds. Köppel, H., Barentzen, H., and Yarkony, D. R., Springer, Heidelberg, 2009, Chapter 1, pp 3-23.

[65] Garcia-Fernandez, P., and Bersuker, I. B., Class of molecular and solid state systems with correlated magnetic and dielectric bistabilities induced by the pseudo Jahn-Teller effect, *Phys. Rev. Lett.* 106 (2011) 246406.

[66] Venkataramani, S., Jana, U., Dommaschk, M., S¨onnichsen, F. D., Tuczek, F., and Herges, R., *Science* 331, 445 (2011).

[67] Ihara, Y., and Ryokichi, T., *Bull. Chem. Soc. Jpn.* 57, 2829 (1984).

[68] Hoffmann, M. M., Darab, J. G., Palmer, B. J., and Fulton, J. L., *J. Phys. Chem. A* 103, 8471 (1999).

[69] Bilc, D. I., R. Orlando, R. Shaltaf, G.-M. Rignanese, J. I˜niguez, and P. Ghosez, Hybrid exchange-correlation functional for accurate prediction of the electronic and structural properties of ferroelectric oxides, *Phys. Rev. B* 77, 165107 (2008).

[70] Berger R. and Tergenius L.-E., Room temperature synthesis and structural characterization on monoclinic $LiCuO_2$ by X-ray and neutron diffraction, *J. Alloys and Comp.* 203, 203 (1994).

[71] Pickardt, J., Paulus, W., Schmalz, M., and Schöllhorn C., Crystal growth and structure refinement of $NaCuO_2$ by X-ray and neutron diffraction, *J. Sol. Sta. Chem.* 89, 308-314 (1990); Singh, D. J., Electronic structure of $NaCuO_2$ *Phys. Rev. B* 49, 1580 (1994).

[72] Owens, F. J., Evidence of a phase transition in Cu-O chains of $LiCuO_2$, *Physica C* 65-69 (1999).

[73] Lee, S., Pirogov, A., Kang, M., et al., Giant magneto-elastic coupling in multiferroic hexagonal manganites, *Nature* 451, 805 (2008).

[74] Novoselov, K.S., Geim, A.K., Morozov, S.V., Jiang, D., Zhang, Y., Dubonos, S.V., Grigorieva, I.V., Firsov, A.A., Electric field effect in atomically thin carbon films, *Science*, 306, 666 (2004).

[75] 2D Materials and Beyond (Synthesis and Doping, Scanned Probe Techniques, 2D Material Devices, Optical Properties of 2D Materials), *Virtual Issue*, American Chemical Society Virtual Collection, 2017.

[76] Das, S., Robinson, J.A., Dubey, M., Terrones, H., and Terrones, M., Beyond graphene: progress in novel two-dimensional materials and van der waals solids, *Annu. Rev. Mater. Res.* 45, 1(2015).

[77] Jariwala, D., Marks, T.J., and Hersam, M.C., Mixed-dimensional van der Waals heterostructures, *Nat. Mater.*, 16, 170 (2017).

[78] T. Low, T., Chaves, A., Calswel, J.D., Kumar, A., Fang, N.X., Avouris, P., Heinz, T.F., Giunea, F., Martin-Moreno, L., and Koppens, F., Polaritons in layered two-dimensional materials, *Nat. Mater.* 16, 182 (2017).

[79] Brar, Victor W., Koltonow, Andrew R., and Huang, Jiaxing, New discoveries and opportunities from two-dimensional materials, *ACS Photon.*, 4, 407 (2017).

[80] Singh, Arunima K., Revard, Benjamin C., Ramanathan, Rohit, Ashton, Michael, Tavazza, Francesca, and Hennig, Richard G., Genetic algorithm prediction of two-dimensional group-IV dioxides for dielectrics, *Phys. Rev. B* 95 (2017) 155426.

[81] Molle, A., Goldberger, J., Houssa, M., Xu, Y., Zhang, S.-C., and Akinwande, D., Buckled two-dimensional Xene sheets, *Nat. Mater.* 16, 163 (2017).

[82] Wang, H., Yuan, H., Hong, S., Lib, Y., and Cui, Y., Physical and chemical tuning of twodimensional transition metal dichalcogenids, *Chem. Soc. Rev.* 44, 2664 (2015).

[83] Ge, Cui-Huan, Li, Hong-Lai, Zhu, Xiao-Li, and Pan, An-Lian, Band gap engineering of atomically thin two-dimensional semiconductors, *Chin. Phys. B* 26, 034208 (2017).

[84] Bersuker, I. B., Modern aspects of the Jahn-Teller effect theory and applications to molecular problems, *Chem. Rev.* 101, 1067 (2001).

[85] Hermoso, W., Ilkhani, A.R., and Bersuker, I. B., Pseudo Jahn-Teller origin of instability of planar configurations of hexa-heterocycles C4N2H4X2 (X = H, F, Cl, Br), *Comput. Theor. Chem.* 1049, 109 (2014); Ilkhani, A.R., Hermoso, W., and Bersuker, I. B., Pseudo Jahn-Teller origin of instability of planar configurations of hexa-heterocycles. Application to compounds with C_4X_2 skeletons (X = O, S, Se, Te), *Chem. Phys.* 460, 75 (2015).

[86] Y. Liu, Y., Bersuker, I. B., Zou, W., and Boggs, J.E., Pseudo Jahn-Teller versus Renner-Teller effects in instability of linear molecules, *Chem. Phys.* 376, 30 (2010).

[87] Sergeeva, Alina P., and Boldyrev, Alexander I., Flattening a puckered pentasilacyclopentadienide ring by suppression of the pseudo Jahn-Teller effect, *Organometallics* 29, 3951 (2010).

[88] Pokhodnya, Konstantin, Anderson, Kenneth, Kilina, Svetlana, and Boudjouk, Philip, Toward the mechanism of perchlorinated

cyclopentasilane (Si5Cl10) ring flattening in the $[Si_5Cl_{10}\text{-}2Cl]^{2-}$ dianion, *J. Phys. Chem. A* 121, 3494 (2017).

[89] Pokhodnya, K., Olson, C., Dai, X., Schulz, D.L., Boudjouk, P., Sergeeva, A.P., and Boldyrev, A., Flattening a puckered cyclohexasilane ring by suppression of the pseudo-Jahn–Teller effect, *J. Chem. Phys.* 134, 014105 (2011).

[90] Ivanov, Alexander S., Bozhenko, Konstantin V., and Boldyrev, Alexander I., On the suppression mechanism of the pseudo-Jahn–Teller effect in middle E_6 (E = P, As, Sb) rings of triple-Decker sandwich complexes, *Inorg. Chem.* 51, 8868 (2012).

[91] Ilkhani, Ali R., Gorinchoy, Natalia N., and Bersuker, Isaac B., Pseudo Jahn-Teller effect in distortion and restoration of planar configurations of tetra-heterocyclic 1,2-diazetes $C_2N_2E_4$, E = H, F, Cl, Br, *Chem. Phys.* 460, 106 (2015).

[92] Liu, Yang, Wang, Ya, and Bersuker, I. B., Geometry, electronic structure, and pseudo Jahn-Teller effect in tetrasilacyclobutadiene analogues, *Nature. Sci. Rep.* 6, 23315 (2016); Liu, Yang, Bersuker, I. B., Garcia-Fernandez, P., and Boggs, J.E., Pseudo Jahn-Teller origin of nonplanarity and rectangular ring structure of tetra-fluorocyclobutadiene, *J. Phys. Chem. A* 116, 7564 (2012); Liu, Yang, Bersuker, I. B., and Boggs, J.E., Pseudo JahnTeller origin of puckering in $C_4H_4^{2+}$, $Si_4H_4^{2+}$, and $C_4F_4^{2+}$ dications, *Chem. Phys.* 417, 26 (2013).

[93] Jose, Deepthi, and Datta, Ayan, Understanding of the buckling distortions in silicene, *J. Phys. Chem. C* 116, 24639 (2012).

[94] Jose, Deepthi, and Datta, Ayan, Structures and chemical properties of silicene: unlike graphene, *Acc. Chem. Res.* 47, 593 (2014).

[95] Chowdhury, Chandra, Jahiruddin, Sheik, and Datta, Ayan, Pseudo-Jahn–Teller distortion in two-dimensional phosphorus: origin of black and blue phases of phosphorene and band gap modulation by molecular charge transfer, *J. Phys. Chem. Lett.* 7, 1288 (2016).

[96] Qiao, Jingsi, Zhou, Linwei, and Ji, Wei, Geometric stability and electronic structure of infinite and finite phosphorus atomic chains, *Chin. Phys. B* 26, 036803 (2017).

[97] Li, Likai, Kim, Jonghwan, Jin, Chenhao, Ye, Guo Jun, Qiu, Diana Y., Jornada, Felipe H. Da, Shi, Zhiwen, Chen, Long, Zhang, Zuocheng, Yang, Fangyuan, Watanab, Kenji, Taniguchi, Takashi, Ren, Wencai, Louie, Steven G., Chen, XianHui, Zhang, Yuanbo, and Wang, Feng, Direct observation of the layer-dependent electronic structure in phosphorene, *Nat. Nanotechnol.* 12, 21 (2017).

[98] Nijamudheen, A., Bhattacharjee, Rameswar, Choudhury, Snehashis, and Datta, Ayan, Electronic and chemical properties of germanene: the crucial role of buckling, *J. Phys. Chem. C* 119, 3802 (2015).

[99] Pratik, Saied Md., and Datta, Ayan, 1,4-Dithiine-Puckered in the gas phase but planar in crystals: role of cooperativity, *J. Phys. Chem. C* 119, 15770 (2015).

[100] Gorinchoy, N.N., and Bersuker, I. B., Pseudo Jahn-Teller effect in control and rationalization of chemical transformations in two-dimensional compounds, *J. Phys. Conf. Ser.* 833, 012010 (2017).

[101] Chou, J.H., Rauchfuss, T.B., and Szczepura, L.F., *J. Am. Chem. Soc.* 1998, 120 (1805).

[102] Beck, J., Weber, J., Mukhopadhyay, A.B., and Dolg, M., Bent and planar molecules in polymorphs of the tricyclic carbon sulfide C_6S_8, *New J. Chem.* 29, 465 (2005).

[103] Ivanov, Alexander S., Miller, Evan, Boldyrev, Alexander I., Kameoka, Yuichiro, Sato, Tohru, and Tanaka, Kazuyoshi, Pseudo Jahn–Teller origin of buckling distortions in two-dimensional triazine-based graphitic carbon nitride (g-C3N4) sheets, *J. Phys. Chem. C* 119, 12008 (2015).

[104] Polinger, V., Garcia-Fernandez, P., and Bersuker, I. B., Pseudo Jahn-Teller origin of ferroelectric instability in BaTiO3 type perovskites. The Green's function approach and beyond, *Physica B: Cond. Matter*, 457, 296 (2015).

[105] Peierls, R.E., *Quantum Theory of Solids*, Clarendon, Oxford, 1955.

BIOGRAPHICAL SKETCH

Isaac B. Bersuker

Affiliation: The University of Texas at Austin, USA

Education: Bachelor in Science from Kishinev University, Moldova, in 1952; PhD in 1957 and Doctor Habilitat in 1964 from Leningrad University (USSR).

Research and Professional Experience:

In his PhD dissertation I. B. Bersuker revealed the influence of core polarization on optical transition in atoms (a 1957 paper is still cited in the literature). His main interests afterward were in the Jahn-Teller (JT), pseudo JT (PJT), and Renner-Teller effect theory, and a large variety of their applications. Among his main achievements in this field, he predicted the tunneling splitting in JT systems, proved that the JT and PJT effects are the only source of structural symmetry breaking in polyatomic systems, identified the hidden JT and PJT effects, predicted the coexistence of trigonal and tetragonal JT distortions in T-term JT effect, created the vibronic (PJT) theory of ferroelectricity, discovered the role of spin in spontaneous polarization of crystals and new classes of multiferroics with magnetic-ferroelectric crossover, related vibronic energy level degeneracies to multi-conical intersections and Berry phase, revealed the role of JT and PJT effects in mixed-valence compounds, worked out methods of evaluation JT effect parameters from ultrasonic experiments.

For materials with JT and PJT effects, revealed novel specific properties, including magnetic-dielectric bistability, origin of buckling in 2D systems, giant flexoelectricity, permittivity, electrostriction, and polar nanoregions in paraelectric phases of ferroelectric perovskites.

In a more chemical oriented series of publications, he worked out methods of calculation of PJT induced molecular instabilities, introduced the plasticity effect in stereochemistry, revealed the origin of puckering in molecular systems, showed that bending of linear molecules is induced by the PJT effect (not Renner-Teller effect), worked out a variety of other chemical applications of the JT and PJT effects, including the theory vibronic activation by transition metals in coordination compounds, biological systems, and mixed-valence clusters, explored the role of the PJT effect in hemoglobin oxygenation, revealed the chemical-structural origin of musk fragrance activity, suggested and worked out a method of combined quantum/classical (QM/MM) modeling for large organometallic and metallo-biochemical systems with charge transfer between the QM and MM fragments, and a more "application oriented" Electron-Conformational method of pharmacophore identification and bioactivity prediction in drug design and toxicology.

Professional Appointments: Head of Laboratory of Quantun Chemistry, Academy of Sciences of Moldova, from 1969 to 1993, Senior researcher and Professor at the Department of Chemistry, the University of Texas at Austin, from 1993.

Honors: Medal of Honor (2004, Moldova), Full Member of the Academy of Science (Moldova, elected in 1988), State Prize Laureate (1980, Moldova), USSR state registered Scientific Discovery N 202 (1979), a series of other medals.

Publications from the Last 3 Years:

1) Bersuker, I. B., Spontaneous Symmetry Breaking in Matter Induced by Degeneracy and Pseudodegeneracy, in *Advances in Chemical Physics*, Vol. 160, S. Rice and R. Dinner, eds., Wiley, New York, 2016, pp159-208.
2) Liu, Yang, Ya Wang, and Isaac B. Bersuker, Geometry, Electronic Structure, and Pseudo Jahn-Teller Effect in Tetrasilacyclobutadiene Analogues, *Nature - Scientific Reports*, 6, 23315 (2016); doi: 10.1038/srep23315 (2016).
3) Zhevstovskikh, Irina V., Vladimir V. Gudkov, Maksim N. Sarychev, Sergei Zherlitsyn, Shadi Yasin, Isaac B. Bersuker, Nikita S. Averkiev,

Kirill A. Baryshnikov, Andrey M. Monakhov, Yuriy V. Korostelin, Magnetic Field Induced Relaxation Attenuation of Ultrasound by Jahn–Teller Centers: Application to ZnSe:Cr^{2+}, *Appl. Magn. Reson.,* 47, 685–692 (2016); DOI 10.1007/s00723-016-0765-9.

4) Zhevstovskikh, I. V., I. B. Bersuker, V. V. Gudkov, N. S. Averkiev, M. N. Sarychev, S. Zherlitsyn, S. Yasin, G. S. Shakurov, V. A. Ulanov, V. T. Surikov, Numerical adiabatic potentials of orthorhombic Jahn-Teller effects retrieved from ultrasonic experiments. Application to the SrF_2:Cr crystal, *J. Applied Physics* 119, 225108 (2016).

5) Bersuker, I. B., The Jahn-Teller and Pseudo Jahn-Teller Effect in Materials Science, *J. Phys. Conf. Series,* 833, 012001 (2017); doi:10.1088/1742-6596/833/1/012001 393.

6) Averkiev, N. S., I. B. Bersuker, V. V.Gudkov, I. V. Zhevstovskikh, M. N.Sarychev, S. Z. herlitsyn, Sh. Yasin, Y. V. Korostelin, V. T. Surikov, Elastic Moduli of Ultrasound Attenuation in Cadmium Selenide doped with Chromium, *J. Appl. Math. Phys.*, 5, 26-30 (2017); http://dx.doi.org/10.4236/jamp.2017.51004.

7) Polinger V., and I. B. Bersuker, Pseudo Jahn-Teller effect in permittivity of ferroelectric perovskites, *J. Phys. Conf. Series,* 833, 012012 (2017); doi :10.1088/1742-6596/833/1/012012.

8) Averkiev, N. S., I. B. Bersuker, V. V. Gudkov, I. V. Zhevstovskikh, M. N. Sarychev, S. Zherllitsyn, S. Yasin, G. S. Shakurov, V. A. Ulanov and V. T. Surikov, Manifestation of the Jahn-Teller effect in elastic moduli of strontium fluorite crystals doped with chromium ions, *J. Phys. Conf. Series* 833, 012003 (2017); doi :10.1088/1742-6596/833/1/012003.

9) Gorinchoy N. N., and I. B. Bersuker, Pseudo Jahn-Teller effect in control and rationalization of chemical transformations in two-dimensional compounds, *J. Phys. Conf. Series,* 833, 012010 (2017); doi :10.1088/1742-6596/833/1/012010.

10) Hizhnyakov, V., I. B. Bersuker, and P. Garcia-Fernandez, Eds., Vibronic Coupling and Electron-Phonon Interactions in Molecules

and Crystals: XXIII International Symposium on the Jahn-Teller Effect, Foreword, *J. Phys. Conf. Series*, 833, 011001 (2017).

11) Averkiev, N. S., I. B. Bersuker, V. V. Gudkov, I. V. Zhevstovskikh, K. A. Baryshnikov, M. N. Sarychev, S. Zherlitsyn, S. Yasin, and Yu.V. Korostelin, Magnetic field induced tunneling and relaxation between orthogonal configurations in solids and molecular systems, *Phys. Rev. B*, 96, 094431 (2017).

12) Averkiev, Nikita S., Isaac B. Bersuker, Vladimir V. Gudkov,, Irina V. Zhevstovskikh, Maksim N. Sarychev, Sergei Zherlitsyn, Shadi Yasin, Gilman S. Shakurov, Vladimir A. Ulanov, and Vladimir T. Surikov, Acoustic Properties of Crystals with Jahn–Teller Impurities: Elastic Moduli and Relaxation Time. Application to $SrF_2:Cr^{2+}$, *J. Phys. Soc. Jpn.* 86, 114604 (2017); DOI: 10.7566/JPSJ.86.114604.

13) Bersuker, I. B., Manipulation of Structure and Properties of Two-Dimensional System Employing the Pseudo Jahn-Teller effect, *FlatChem,* 6, 11–27 (2017); DOI: 10.1016/j.flatc.2017.10.001.

14) Bersuker, I. B., The Vibronic (Pseudo Jahn-Teller) Theory of Ferroelectricity. Novel Aspects and Applications, *Ferroelectrics*, 536, 1-59 (2018); https://doi.org/10.1080/00150193.2018.1528919.

15) Sarychev, M. N., Irina V. Zhevstovskikh, Nikita S. Averkiev, Isaac B. Bersuker, Vladimir V. Gudkov, V. T. Surikov, Evaluation of the parameters of the Jahn-Teller effect in impurity centers in crystals. Application to the crystal $ZnSe:Ni^{2+}$, *Russian Solid State Physics*, 61(2), 319-324 (2018).

16) Polinger, V. and I. B. Bersuker, Origin of Polar Nanoregions and Relaxor properties of ferroelectrics, *Phys. Rev. B*, 98, 214102 (2018). DOI:10.1103/PhysRevB.98.214102.

17) Averkiev, N. S., I. B. Bersuker, V. V. Gudkov, I. V. Zhevstovskikh, M. N. Sarychev, S. Zherlitsyn, S. Yasin, Yu. V. Korostelin and V. T. Surikov, Jahn-Teller effect problems via ultrasonic experiments. Application to the impurity crystal, CdSe:Cr, *J. Phys. Conf. Series,*

1148 (2018) 012008; doi:10.1088/Journal of 1742-6596/1148/1/012008.

18) Gorinchoy, N. N., I. Arsene, and I. B. Bersuker, Buckybowl Structure of Sumanenes and Distortions of Thiophenes Induced by the Pseudo Jahn-Teller Effect, *J. Phys. Conf. Series,* 1148 (2018) 012005; doi:10.1088/1742-6596/1148/1/012005.

In: Recent Studies in Materials Science
Editor: Patrick R. Lind
ISBN: 978-1-53615-270-8

Chapter 2

CONVECTIVE EFFECTS AND TRAVELING WAVES IN TRANSPARENT OXIDE MATERIALS PROCESSED WITH THE FLOATING ZONE TECHNIQUE

Marcello Lappa*

Department of Mechanical and Aerospace Engineering,
University of Strathclyde, Glasgow, Scotland, UK

ABSTRACT

Zone melting (or zone refining or floating zone process, FZ) is a group of similar methods, specifically conceived for the purification of crystals, in which thermally-driven flows of both gravitational and surface-tension natures are typically produced when the considered material is processed. Since the melt never comes into contact with anything but vacuum (or inert gases), there are no contaminants that the melt may incorporate. Even though compounds with higher purity and improved quality can be obtained with this technique, a typical drawback is represented by the

* Corresponding Author Email: Marcello.lappa@strath.ac.uk.

defects potentially induced in the crystalline structure by the unavoidable convection emerging in the fluid phase. In the present chapter, special attention is paid to a specific category of materials known as transparent oxides. A range of conditions is explored, differing in the dominant effect (buoyancy or Marangoni flow), the thermal conditions (heating being provided along the radial or axial direction) and the relative direction of gravity and applied temperature gradient. The hallmark of the entire chapter is our commitment to identify situations in which "waves" are produced and provide a systematic classification of such convective instabilities together with a description of related features based on advanced numerical simulations.

1. Introduction

Relevant examples pertaining to the category of oxide transparent materials include (but are not limited to): yttrium aluminum ($Y_3Al_5O_{12}$), yttrium orthoaluminate ($YAlO_3$), spinel ($MgAl_2O_4$), sapphire (Al_2O_3), lithium tantalite ($LiTaO_3$) and gadolinium gallium garnet ($Gd_3Ga_5O_{12}$). These compounds constitute a basis for a variety of advanced industrial applications. Indeed, they are used for solid state lasers, and as magnetic bubble device substrates, insulating layers for semiconductors, or monolithic crystal filters. Moreover, the so-called transparent conductive oxides (TCOs), known for their unusual ability to possess transparent and conducting properties at the same time (Ohta and Hosono, 2004), are typically employed in flat panel displays (such as LCDs, i.e., liquid crystal displays), electroluminescent devices (e.g., organic light emitting diodes - OLEDs), solar cells and power saving opto-electrical circuitries (Stadler, 2012). Notably, the transparency property also applies to the related molten state, which is typically established at relatively high temperatures (e.g., the melting points of sapphire and yttrium aluminum are $\cong$ 2300 K and 2960 K, respectively).

The kinematic viscosity (ν) and thermal diffusivity (α) of all these melts have, in general, a comparable order of magnitude (with the former being slightly larger than the latter). This behavior formally puts these substances in the general class of *high-Pr liquids*, where $Pr = \nu/\alpha$ is the so-called fluid

Prandtl number. This parameter generally ranges in the interval O(1) ≤ Pr ≤ O(10) and this might be regarded as a clear distinguishing mark with respect to other transparent insulating materials (e.g., SiO_2 glasses for which Pr = O(10^3)), organic substances such as molten hexatriacontane ($C_{36}H_{74}$, Pr = 65) and molten tetracosane ($C_{24}H_{50}$, Pr = 49) or opaque liquid metals and semiconductor (or superconductor) melts, for which Pr<<1.

From a purely industrial standpoint, transparent oxides are generally crystallized using different methods such as vapor-phase or liquid-phase epitaxial techniques or other processes which come under the heading of 'crystal-growth-from-the-melt' (relevant examples being represented by the Horizontal or vertical Bridgman technique, the Czochralski (CZ) method or the so-called Floating zone (FZ) process, Hurle, 1994; Lappa, 2009). A related key aspect is represented by the ability to transform the considered transparent compound, initially in a polycrystalline state, into a *single-crystal* while keeping the formation of impurities and defects *at minimum*. In such a context, many advantages are provided by the *FZ process* given its (known) matchless ability to reduce melt contamination potentially induced by the interaction of the high-temperature melt with external walls or parts.

The FZ process starts from a cylindrical rod of the considered polycrystalline material to be melted in a controlled environment (an inert gas or vacuum) via the application of localized heating to its surface. Such heating is generally provided in order to preserve the initial symmetry of the specimen, i.e., by means of a coaxial ring heater or an optical furnace relying on curved (circular) mirrors. The rod does not hold a stationary position with respect to the heating device. Rather relative motion is established between the material and the limited region subjected to localized heating. In this way, an ongoing process is maintained where a new portion of the polycrystalline bar undergoes melting while the liquid re-solidifies as a single crystal (the 'seed') at the other side of the liquid zone. This means that, at any instant, the entire amount of molten material is suspended (in the form of a 'liquid bridge') between the initially impure solid and the seed (which illustrates why these crystallization processes come under the common heading of "Floating Zone" approach).

The intrinsic advantages become readily evident if one considers that the liquid is never in contact with a container or a crucible, which may dissolve thereby altering its composition. As a result, crystals with higher purity can be produced (see, e.g., Benz, 1990). These advantages have made these methods increasingly popular for the production of a wide range of materials, including metals, semiconductors (Cröll et al., 1998), high-temperature superconductors, new magnetic materials and oxides (Saurat and Revcolevsch, 1971; Shindo et al., 1979; Shindo, 1980; Balbashov and Egorov, 1981; Kimura and Kitamura, 1992; Revcolevschi and Jegoudez, 1997; Moest et al., 1998). In particular, transparent oxides typically processed with the FZ are substances such as β-Ga_2O_3 (Villora et al., 2004) or TiO_2 (Higuchi and Kodaira, 1992; Watauchi et al., 2012), and more complex compounds such as $CuGeO_3$ (Revcolevschi and Jegoudez, 1997) and $Y_3Fe_5O_{12}$ (Balbashov et al., 1975; Shindo et al., 1979; Balbashov and Egorov, 1981). The interested reader is referred to the works by Dabkowska and Gaulin (2003) and Dabkowska and Dabkowski (2010) for an exhaustive review of materials processed with the FZ technique.

Here, we concentrate on a specific aspect, that is, the fluid convection occurring in the molten zone (the aforementioned "liquid bridge"), given its known ability to induce undesired (often unavoidable) compositional and structural defects in the final crystals (Muller, 1988).

Fluid motion obviously arises as a consequence of the significant thermal gradients being present in the melt as a result of the temperature difference between relatively cold regions (where the material is solid) and those where heating is applied.

Remarkably, these gradients can support two concurrent mechanisms driving fluid motion, the first being obviously represented by the well-known dependence of the density of liquids on temperature (which, by enabling buoyancy effects, can lead to convection of thermogravitational nature). The second relates directly to the existence of a free surface in the considered process, namely the interface that separates the melt region from the external environment (be it an inert gas or vacuum). The stable existence of this surface is allowed by surface tension (σ) that, by counteracting gravity, prevents the liquid from being pulled down. As the surface tension depends on temperature (the higher the temperature, the smaller σ), any temperature gradient established along the interface is directly turned into a gradient of surface tension forcing the fluid to move from warmer surface regions towards colder ones. The related mechanism is generally known as thermocapillary (or Marangoni) convection.

Even though their relative strength can vary depending essentially on the considered fluid and the effective extension of the liquid zone, in typical industrial applications these two forms of convection are always simultaneously present (Lappa, 2009, 2012). Their relative importance or weight can also have a remarkable impact on the so-called *hierarchy of bifurcations*, i.e., the specific path followed by these systems when the temperature difference (ΔT) is increased. A rise in the ΔT typically results in the flow undergoing *various symmetry breaking phenomena* and in *the onset of time-dependence*. The precise sequence of events, however, does depend on various parameters and on the dominant driving force.

The present chapter is entirely devoted to a focused review of these modes of convection in physical domains with cylindrical symmetry (assumed to mimic the real FZ process), together with a critical analysis of very recent findings.

Given the impressive amount of literature existing on these subjects, we limit ourselves to discussing fluids with Pr $\geq$ O(1) representative of transparent oxides, the reader being referred to Lappa (2007) for an analogous treatment concerning liquid metals and semiconductor melts.

Emphasis is also given to aspects relating to the "pattern formation", particularly for the circumstances in which thermocapillary and thermogravitational convection, in disjoint or combined form, can produce undesired *waves traveling* in the considered fluid system. With the descend of scale, localized defects produced by such waves in the material start to be dominating with respect to other types of defects induced by large-scale flow; complex interplay of these together contributes to significant manufacturing challenges (Hurle, 1994).

In order to place all these phenomena in a proper theoretical context, we provide relevant details about the overarching fluid-dynamic equations, typical mathematical models used for the FZ technique, the related boundary conditions and underlying assumptions. Along the same lines, the discussion is supported by a variety of numerical results to show the considered dynamics with a level of detail that is hardly achievable by experimental analysis and/or direct flow visualization.

2. MATHEMATICAL MODELS

2.1. Governing Equations

Referring velocity and temperature to the scales α/L and ΔT, respectively and scaling all distances on L, where L is a characteristic length for the considered problem and α is the liquid thermal diffusivity, the overarching balance equations for mass, momentum and energy transport in the molten material can be cast in condensed form as :

$$\underline{\nabla}\cdot\underline{V}=0 \tag{1}$$

$$\frac{\partial\underline{V}}{\partial t}=-\underline{\nabla}p-\underline{\nabla}\cdot\left[\underline{V}\underline{V}\right]+\Pr\nabla^2\underline{V}-\Pr Ra\ T\ \underline{i}_g \tag{2}$$

$$\frac{\partial T}{\partial t}+\underline{\nabla}\cdot\left[\underline{V}T\right]=\nabla^2 T \tag{3}$$

where $\underline{V}$, T and p are the nondimensional velocity, temperature and pressure, respectively and $\underline{i}_g$ is the unit vector along the direction of gravity. The parameter $Ra=\dfrac{g\beta_T\Delta TL^3}{\nu\alpha}$, known as the Rayleigh number, essentially results from the application of the so-called Boussinesq approximation (Lappa, 2009) to the term accounting for the buoyancy force in the momentum equation (where a linear relationship is assumed between density and temperature via the thermal expansion coefficient β_T). The Rayleigh number formally measures the relative importance of buoyancy and viscous forces.

Closure of the problem from a mathematical point of view, however, also requires that proper boundary conditions are used to complement eqs. (1)-(3). This means that a more complete mathematical framework must be defined, i.e., an abstraction of physical reality sufficiently simple to allow the treatment of the considered problem with a limited number of parameters and, at the same time, able to retain the physical mechanisms which are thought to be significant or of specific interest. Along these lines, the next section is entirely devoted to a presentation of the mathematical models which have enjoyed widespread use over recent years for the FZ process.

2.2. Geometrical Models of the FZ Technique

As explained above, the approach is to consider surrogates of the FZ that incorporate the presumed influential (behavior-determining) process(es). Irrespective of the approach used to define them, these configurations are approximate, i.e., a series of choices have to be made in their formulation, which may be dictated by convenience or simplicity or by the need to concentrate on some specific aspects.

In general, these distinct models can be delineated by retaining or disregarding the possible symmetries of the real FZ along different spatial directions. Prior to embarking in the description of them, we wish to remark that the existence of these symmetries implies the possibility of symmetry breaking (a very important concept in fluid-dynamics, given its deep connections with the notion of flow instability and bifurcation, which will be treated later in this chapter).

The simplest approximation for the FZ process is represented by a liquid zone with cylindrical interface held between two disks at different temperature, ideally assumed to model the heat and mass transfer processes occurring between the equatorial plane of a real floating zone (where heating is applied) and one of the two horizontal solid surfaces (where liquefaction or solidification takes place depending on the considered portion of the rod).

As this model accounts only for half size of the effective liquid region, it is generally known as the 'half zone' (Figure 1a). As the main direction along which the temperature difference is applied is *axial*, for simplicity, the free interface is generally approximated as an adiabatic boundary.

This assumption, however, must necessarily be removed when the more sophisticated 'full zone' is considered (Figure 1b). With this alternate configuration, heat is injected into the liquid through the free surface (along the radial direction, as it occurs in reality). The two ends of the domain perpendicular to the axial direction are therefore set at the same temperature (ideally mimicking the melting/solidification point of the considered material).

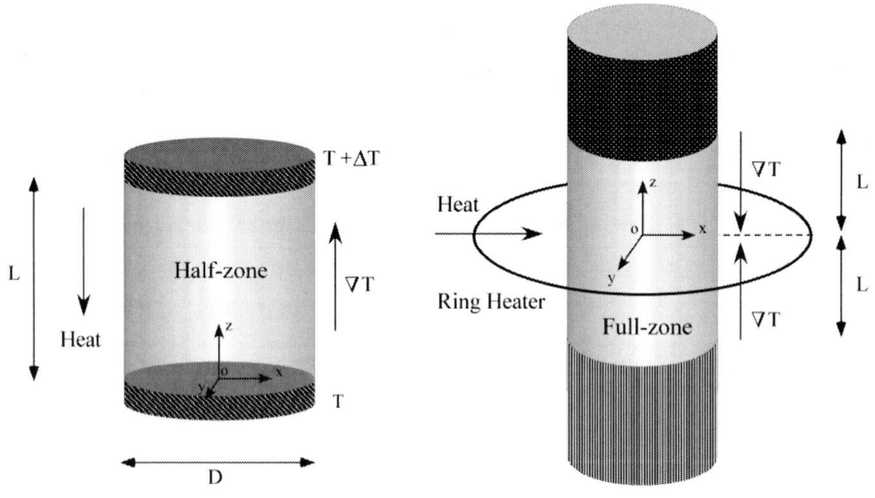

Figure 1. Main models of the FZ process and related boundary conditions: a) the Half Zone, b) the Full Zone.

The aspects, for which the two models can be distinguished, however, are not limited to the direction along which heat is injected into the liquid. As the reader may expect (given the introductory arguments given at the beginning of this section), a substantial difference can be found in the *symmetries* that can be broken by the flow. With the half zone, obviously, any processes that depend on the rupture of symmetry with respect to the vertical direction are excluded.

This degree of freedom, however, is present in the full zone, where the fluid *can cross the equatorial plane*. While a single toroidal roll is established in the half zone as a result of thermally driven convective effects, the full zone typically features two opposite toroidal rolls in the two regions which extend from the equatorial plane towards the solid supports. The ideal mirror symmetry of these two counter-rotating circulation systems with respect to the equatorial plane can be lost, leading to overlying rolls displaying different axial extension or even coalescing.

As a common feature, obviously both models allow symmetry breaking with respect to the azimuthal direction, which explains why the half zone has enjoyed so much attention over the last years as a relevant configuration to investigate the onset and propagation of the aforementioned waves (which in such systems travel essentially in the *azimuthal direction*).

From a purely mathematical standpoint, assuming the two supporting disks to be located at z = 0 and z = 1, respectively (reference length L equal to the distance separating the disks), for the half zone, the relevant thermal boundary conditions can be cast in condensed form as:

$$T(z=0, r, \varphi, t) = T_{Bottom} \quad 0 \le r \le 1/A_H \ ; \ 0 \le \varphi \le 2\pi \tag{4a}$$

$$T(z=1, r, \varphi, t) = T_{Top} \quad 0 \le r \le 1/A_H \ ; \ 0 \le \varphi \le 2\pi \tag{4b}$$

where $A_H = L/D$ is the aspect ratio, φ is the azimuthal angle and $T_{Bottom} = 0$ and $T_{Top} = 1$ or $T_{Bottom} = 1$ and $T_{Top} = 0$ according to whether the liquid bridge is *heated from above or from below*, respectively. At the free interface the condition of adiabaticity simply results in:

$$\frac{\partial T}{\partial r}(z, r = 1/2A_H, \varphi, t) = 0 \tag{5}$$

For the full zone, as explained before, the two ends must be set at the same temperature. Assuming as reference length L the axial distance between the equatorial plane containing the ring heater and one of the supporting disks (see Figure 1b), related thermal conditions read:

$$T(z=-1, r, \varphi, t) = 0 \quad 0 \le r \le 1/A_F \ ; \ 0 \le \varphi \le 2\pi \tag{6a}$$

$$T(z=1, r, \varphi, t) = 0 \quad 0 \le r \le 1/A_F ; \ 0 \le \varphi \le 2\pi \tag{6b}$$

where $A_F = 2L/D = 2A_H$. In this case, a slightly more complex treatment is required for the interface as the condition of no heat flux (eq. (5)) must be replaced by an adequate mathematical expression accounting for the radiative heat transfer from the heating system to the liquid surface. Many models have been proposed in literature to simulate surface heating due to a ring heater (see, e.g., Otani et al., 1988; Rivas and Vazquez-Espi, 2001). Following Lappa (2016), the radiative flux generated by a coaxial ring heater with negligible thickness located at a fixed distance h from the free surface can be modeled in a relatively straightforward way starting from the radiative contribution brought to the overall flux by each infinitesimal portion of the ring. Indeed, taking into account the details shown in Figure 2, this contribution can formally be expressed as:

$$j(z) = \left(\frac{1}{L\lambda}\right)\frac{1}{4\pi}\frac{\xi\overline{Q}}{\delta^2}\cos(\theta) \tag{7a}$$

where $\overline{Q}$ is the overall power consumed by the ring heater, δ represents the distance between two generic points lying respectively on the ring heater (C) and on the interface (P) and θ denotes the angle comprised between the direction PC and the unit vector $\underline{\hat{n}}$ perpendicular to the free surface in P.

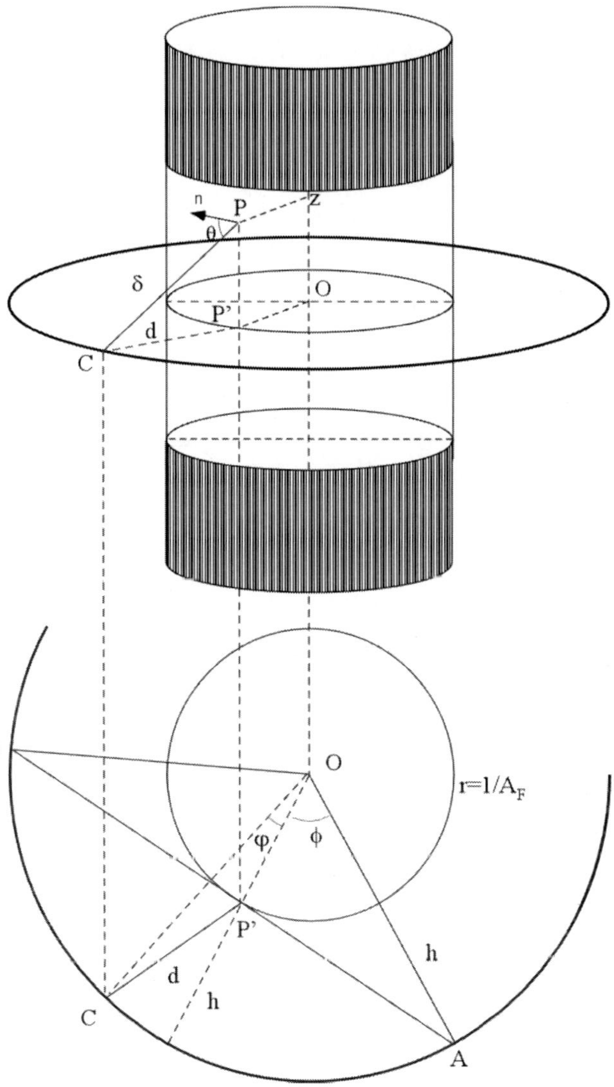

Figure 2. Sketch of the floating zone and related ring heater.

Integration of the infinitesimal contribution along the entire circumferential extension formally gives the effective (local) flux as:

$$J(z) = \frac{1}{4\pi}\left(\frac{\xi \overline{Q}}{L\lambda}\right)\int_{-\tilde{\phi}}^{\tilde{\phi}} \frac{\cos(\theta)}{\delta^2}\, d\varphi \tag{7b}$$

where:

$$\delta^2 = \frac{1}{A_F^2}\left[(1+hA_F)^2 + 1 - 2(1+hA_F)\cos\varphi + z^2 A_F^2\right] \tag{8a}$$

$$\cos(\theta) = \frac{(1+hA_F)\cos\varphi - 1}{\sqrt{\left[(1+hA_F)^2 + 1 - 2(1+hA_F)\cos\varphi + z^2 A_F^2\right]}} \tag{8b}$$

$$\tilde{\phi} = \cos^{-1}\frac{1}{(1+hA_F)} \tag{8c}$$

and ξ is the free-surface emissivity. Therefore, eq. (5) must be replaced by

$$\frac{\partial T}{\partial r}(z, r = 1/A_F, \varphi, t) = J(z) \tag{9}$$

Yet from a mathematical point of view, problem ‘closure’ also requires relevant kinematic conditions along the boundary. For solid surfaces these reduce to the trivial condition of no-velocity (no-slip), i.e., $\underline{V} = 0$.

On the free surface, however, the velocity cannot be imposed directly. Indeed, it must satisfy a dynamic balance between the viscous shear stress in the liquid and the gradient of surface tension. In non-dimensional form such a balance can be cast in compact form as:

$$\left[2(\underline{\nabla}\underline{V})^s\right]\cdot\underline{\hat{n}} = -Ma\left(\underline{\underline{I}} - \underline{\hat{n}}\underline{\hat{n}}\right)\cdot\underline{\nabla}T \tag{10}$$

where $\left(\underline{\nabla V}\right)_o^s = \frac{\underline{\nabla V} + \underline{\nabla V}^T}{2}$ and Ma is the so-called Marangoni number, $Ma = \mathrm{Re}\,\mathrm{Pr} = \frac{\sigma_T \Delta T L}{\mu \alpha}$ (μ being the fluid dynamic viscosity and σ_T the derivative of surface tension with respect to temperature). This number plays for thermocapillary flows the same role that the Rayleigh number has for thermogravitational flows.

Accordingly, a relevant parameter used to characterize the relative importance of surface-tension driven and buoyancy effects (the so-called dynamic Bond number) can be introduced as:

$$B_d = \frac{Ra}{Ma} \tag{11}$$

There are many numerical techniques that can be used to solve directly all these equations (and related boundary conditions) or get useful information from simplified versions of them (e.g., via linear stability analysis). Clarifying the advantages or bottlenecks associated with each different approach is not as straightforward as one would imagine. In this section we will merely touch on this subject, which would otherwise occupy the entire chapter (the reader specifically interested in these aspects may consult the extensive information reported in a companion chapter of this book, Lappa and Ferialdi, 2019).

In particular, we limit ourselves to pointing out that the so-called category of *projection methods* has already proven to be a relevant choice for the simulation of the fluid-dynamic instabilities in floating zones, as witnessed by the many research articles appearing in the literature (Yasuhiro et al., 1997, 1999; Lappa et al., 2000 and 2001; Melnikov et al., 2005; Shevtsova et al., 2001, 2003, 2011; Lappa, 2016). Remarkably, with gained confidence in the validity of these methods, research on these subjects is still progressing and it can still be considered nowadays as a very active line of inquiry.

All the (representative) numerical results described in this chapter have been obtained in the framework of such techniques.

Given the spirit of the present book, conceived to provide readers with applicative aspects and insights into recent findings, such examples are essentially used to support the discussion and explain the 'physics' of the considered phenomena (Sects. 3 and 4). For the above reasons, details about the used meshes, numerical schemes and accuracy are glossed over, the reader being referred directly to the articles cited in the text for additional details.

In particular, most of results are presented for Pr > 10 given the widespread use of such liquids in the literature as surrogate models of oxide melts.

3. The Liquid Bridge

This section is entirely devoted to the popular classical liquid bridge or half zone (hereafter these two terms will be used as synonyms), given its success over the years as an archetypal model for the study of the fundamental properties of Marangoni flows and related hierarchy of bifurcations. As illustrated in Sect. 2.2, this configuration offers several advantages to investigators, especially, the possibility to set a precise temperature difference 'a priori' and assess the system response for increasing values of it, which explains why this model has attracted the attention of research groups with various interests and motivations, leading over the years to the emergence of a common theoretical framework that is generally referred to as the *liquid-bridge problem.* In the following we present a brief account of the historical perspective that led to this framework and its diverse reverberations in the science of fluids and materials.

3.1. Pure Thermocapillary Flow

The mathematical aspects of this problem have been defined in the previous section (see, in particular, Figure 1a and eqs. (4)-(5)). The liquid bridge or half zone can essentially be seen as a differentially heated portion of liquid delimited by a cylindrical free liquid-gas interface. The thermocapillary (Marangoni) flow established in the liquid is initially steady and laminar and reflects the symmetry properties of the hosting domain (i.e., it is axisymmetric). However, when the temperature difference is sufficiently increased (the Marangoni number exceeds a given threshold), the flow becomes three-dimensional. For the case of oxide melts (in particular) and for all high-Pr liquids (in general), the symmetry breaking process is intimately associated with the onset of time-dependence, i.e., the flow becomes 3D and oscillatory at the same time.

This field has reached a certain level of maturity. Experiments conducted in space (e.g., on-board sounding rockets or on the space shuttle, see, e.g., Schwabe et al., 1982; Schwabe and Scharmann, 1984; Monti, 1987; Monti et al., 1995; Chun and Siekmann, 1995; Monti et al., 1998; Schwabe, 2002 and 2005, etc.) have been instrumental in clarifying that this type of instability does not depend on gravity and it should therefore be regarded as an intrinsic 'property' of Marangoni flow (Figure 3).

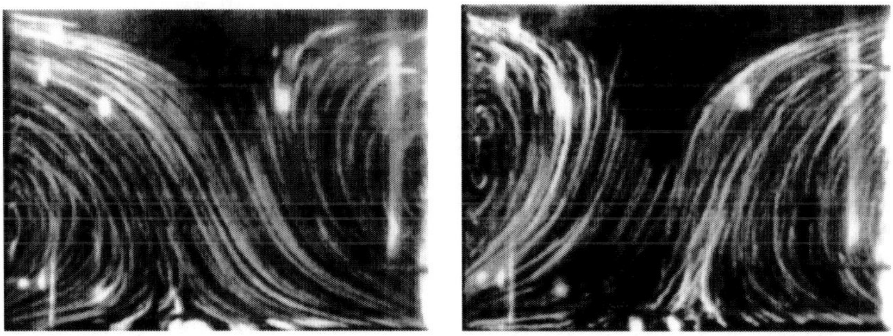

Figure 3. Typical structure of oscillatory Marangoni convection in the meridian plane of a half zone (the motion of tracers is used to visualize the flow; tracers are illuminated by the light generated by a laser diode forming a light sheet perpendicular to the main optical path of a CCD camera; the two snapshots refer to experiments carried out during the Spacelab D2 mission; courtesy of the Microgravity Advanced Research and Support Center).

From an historical standpoint, interestingly, the analysis of this specific problem has progressed through relevant synergy between experimental activities and mathematical arguments based on the application of stability analyses (Neitzel et al., 1992; Kuhlmann and Rath, 1993a; Wanschura et al., 1995; Kuhlmann et al., 1995; Chen et al., 1997). As evident after comparative readings of these studies, starting from considerations based on the axial symmetry of the liquid bridge, it has been shown that 1) the oscillatory disturbances always emerge as *a couple of waves traveling in opposite directions* (namely, a pair of *clockwise and anticlockwise oriented trains of disturbances*) and 2) the specific patterning behavior displayed by the unstable flow in the supercritical regime essentially follows from the *interplay of such waves.*

As originally illustrated by Kuhlmann and Rath (1993b), by modeling each wave as a disturbance with expression $F_{\pm} = B_{\pm}(r,z)\exp\{i[\pm m\varphi - \omega t + G(r,z)]\}$ (where B(r,z) is the wave amplitude, G(r,z) is the phase, m is the wavenumber and $\omega = 2\pi f$ is the angular frequency), and by denoting with η the wave amplitude ratio, the functions expressing wave superposition can be cast in condensed form as:

$$F = 2B(r,z)\cos(m\varphi)\cos(\omega t - G(r,z)) \ (\eta = 1) \tag{12}$$

$$F \cong B(r,z)\ \mathrm{a}(m\varphi)\cos[\mathrm{b}(m\varphi,G) - \omega t] \ (0 < \eta < 1) \tag{13}$$

with $\mathrm{a}(m\varphi) = [\ (1+\eta)^2\cos^2(m\varphi - \varphi_o) + (1-\eta)^2\sin^2(m\varphi - \varphi_o)\ ]^{1/2}$ (14a)

and $\mathrm{b}(m\varphi,G) = \tan^{-1}\left[\dfrac{1-\eta}{1+\eta}\tan(m\varphi - \varphi_o)\right] + G(r,z)$ (14b)

Assembled in this way, eqs. (12) and (13) have a straightforward physical interpretation. As the reader will easily realize, they differ essentially for the location of the group of variables (mφ) at the right hand-side. While in eq. (13) this group is an integral part of the oscillatory term $\cos[\mathrm{b}(m\varphi,G) - \omega t]$, in eq. (12) the dependences on space (φ) and time (t) manifest in disjoint form, namely, $\cos(m\varphi)$ and $\cos(\omega t - G(r,z))$. This apparently innocuous observation has significant consequences on the spatio-temporal dynamics. While in the first case, the maxima and minima of the function *F* at given azimuthal positions can periodically swap their position as soon as cos(ωt-G(r,z)) changes its sign, in the second case the phase of the oscillations depends continuously on φ (which indicates that maxima and minima can continuously travel along the azimuthal direction).

Put differently, according to such arguments, while the superposition of two counter-propagating waves with the same amplitude should simply result in a pattern where the disturbance *nodes occupy fixed positions in space* (a waveform with disturbances growing and shrinking in time at fixed locations in the azimuthal direction), different amplitudes should lead to a pattern whose hallmark is the *visible rotation of features about the symmetry axis*. Using typical nomenclature introduced in the relevant literature, these two possible spatiotemporal modes of convection (waveforms) can also be simply referred to as: the *standing wave* (featured by disturbance nodes pulsating at fixed azimuthal positions, also known as *pulsating pattern*) and a *rotating pattern* (with disturbances traveling circumferentially, for simplicity generally called *traveling wave*).

Continuing with historical developments, it is worth mentioning that the existence of these states has been confirmed by experiments conducted both in space and on the ground (see, e.g., Velten et al., 1991; Frank and Schwabe, 1997; Savino et al., 2001; Schwabe, 2002). In facts, these spatio-temporal behaviors have been directly observed through visualization of tracer particles dispersed in the fluid (through one of the supporting disks made of a material transparent to visible light), or they have indirectly been detected *through analysis of the temperature signals provided by thermocouples at fixed positions* evenly distributed along the circumferential direction.

Following up on this point, the related rationale can directly be gathered from eqs. (12) and (13). As predicted by the first of these two equations, when the flow is a standing wave, all the signals measured by thermocouples located at the same axial and radial coordinates (with different azimuthal positions) must be *in phase or in phase opposition* (put simply, only two values of phase shift are possible, namely $\Delta G = 0$ or $\Delta G = \pi$). For the traveling wave, the set of allowed phase shifts is *continuous* with ΔG (as illustrated, e.g., by Lappa et al., 2001, the phase shift is linearly proportional to the angular distance separating the thermocouples, provided such a distance $\Delta\varphi$ is shorter than π/m).

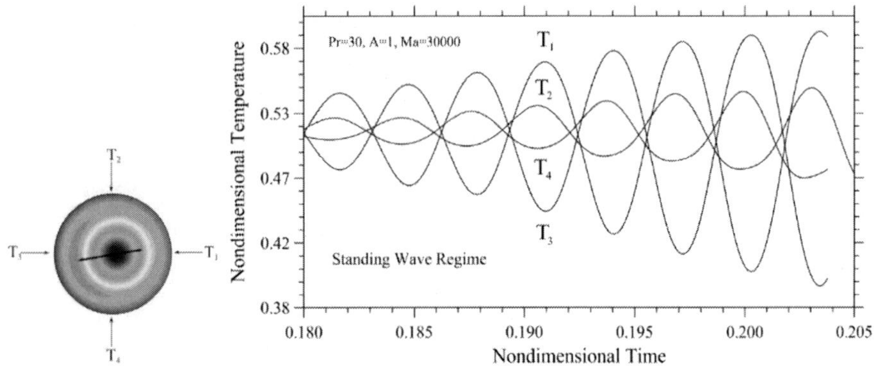

a)

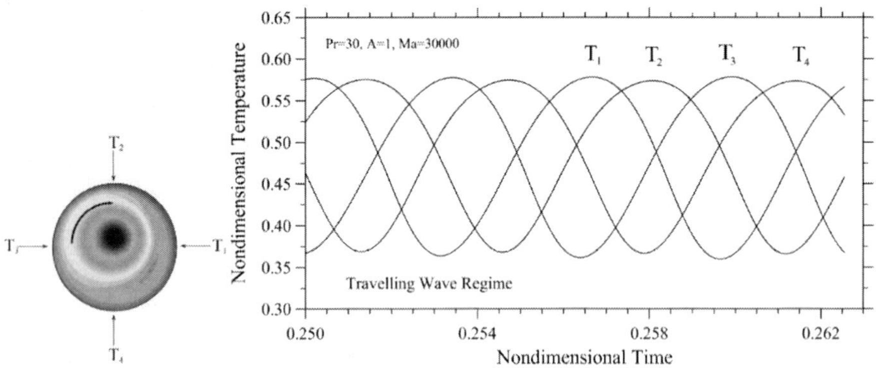

b)

Figure 4. Snapshot of temperature distribution in a section perpendicular to the liquid bridge axis (left view, red and blue colors correspond to regions of relatively hot and cold fluid, respectively) and related temperature signals (right view) provided by four numerical probes periodically positioned at different azimuthal stations along the entire circumferential extension ($A_H = 1$, Pr = 30, Ma = $3x10^4$, m = 1): a) pulsating temperature pattern (standing wave), b) rotating temperature pattern (traveling wave).

In parallel to experiments, numerical simulations have also been instrumental in leading to an exhaustive characterization of these dynamics (see, e.g., Lappa, 1995; Yasuhiro et al., 1997, 1999; Tang et al., 1997; Bazzi et al., 1999; Zeng et al. 1999a,b; 2001b; Leypoldt et al., 2000; Melnikov et al., 2005; Shevtsova et al., 2001, 2003, 2011).

As an example, Figure 4 shows the case for which four thermocouples equally spaced along 360° are used to compare the temperature signals relating to a flow with azimuthal wavenumber m = 1.

It can be seen that in this case the pattern in a cross section perpendicular to the liquid bridge is characterized by the presence of an inner region of colder fluid (this being a typical signature of Marangoni flow in liquid bridges). For m = 1 this region has approximately a circular shape and takes an eccentric position with respect to the point corresponding to the axis of the liquid bridge.

When the flow behaves as a standing wave, such a cold region undergoes back and forth motion along a fixed direction (highlighted by the double black arrow in Figure 4a). As a result the signals measured by the thermocouples display the behavior shown in the underlying plot, i.e., their peaks and valleys align at fixed time stations (in the figure, the valley measured by T_3 and the peak measured by T_1 correspond to the left-most displacement of the internal cold region and vice versa).

For the traveling wave, as evident in Figure 4b, all the peaks and valleys are aligned along the horizontal direction and the temporal distance between two consecutive peaks or valleys is always the same. This indicates that there is no preferred position for the disturbance nodes along the circumferential extension of the liquid bridge and that the center of the aforementioned inner eccentric region simply describes in space a circle, i.e., it travels continuously along the azimuthal direction (in this case the valley measured by each thermocouple corresponds to the instant at which the distance between the considered probe and the inner eccentric region attains a minimum).

Besides temporal features, some differences can also be seen in the spatial morphology taken by the isosurfaces of the azimuthal velocity component at a given instant. As shown in Figure 5a yet for the mode m = 1, for the standing wave, the isosurfaces located in proximity to the free surface look like two opposite "pillows" (corresponding to positive and negative values of the azimuthal velocity, respectively), each confined to one of the two halves of the liquid bridge that would be obtained by cutting it with a plane perpendicular to its bases and containing its axis. As evident in Figure 5b, however, such a 'mental divisibility' is no longer applicable to the traveling wave; when the disturbances rotate, no plane can be identified satisfying such a property regardless of the considered instant. The two pillows are inclined with respect to the axial direction and oriented in such a way that regions of the surface exist where the velocity component close to the bottom disk is positive while it has the opposite sign close to the top (and vice versa).

Additional examples about the typical behavior of standing and traveling waves have been reported in Figures 6 and 7 for a different value of the azimuthal wavenumber.

As the reader will easily realize by inspecting these figures, a general description in terms of eqs. (12)-(14) would still be possible provided m = 1 is replaced with m = 2.

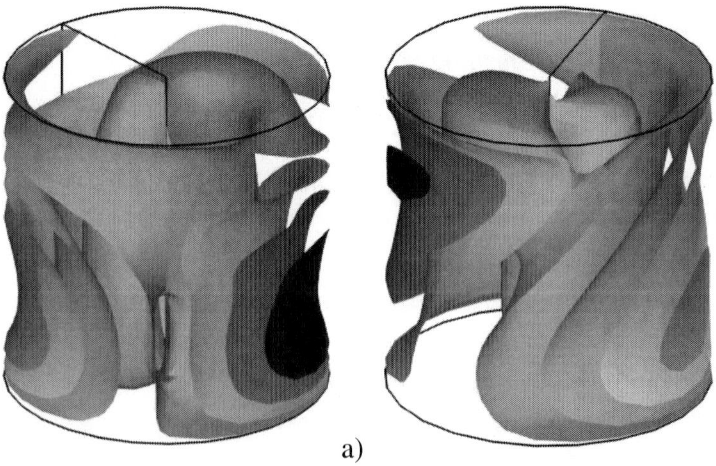

a) b)

Figure 5. Isosurfaces of azimuthal velocity component for the same conditions shown in Figure 4a and 4b, respectively (red and blue colors correspond to regions of positive and negative azimuthal velocity, respectively).

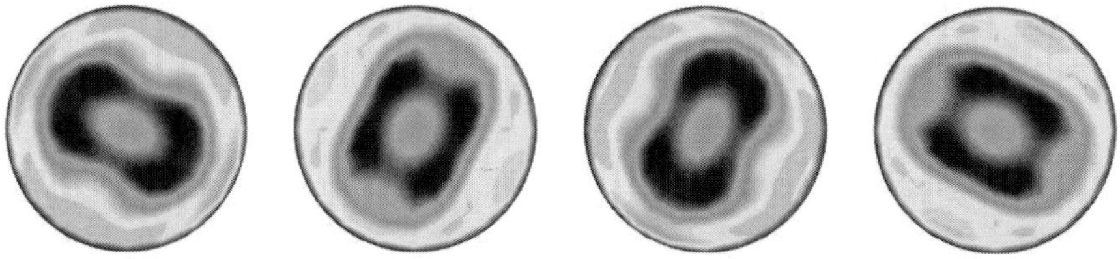

Figure 6. Snapshot of temperature distribution in a section perpendicular to the liquid bridge axis ($A_H = 0.4$, $Pr = 30$, $Ma = 3.6x10^4$, $m = 2$, Standing Wave, red and blue colors correspond to regions of relatively hot and cold fluid, respectively).

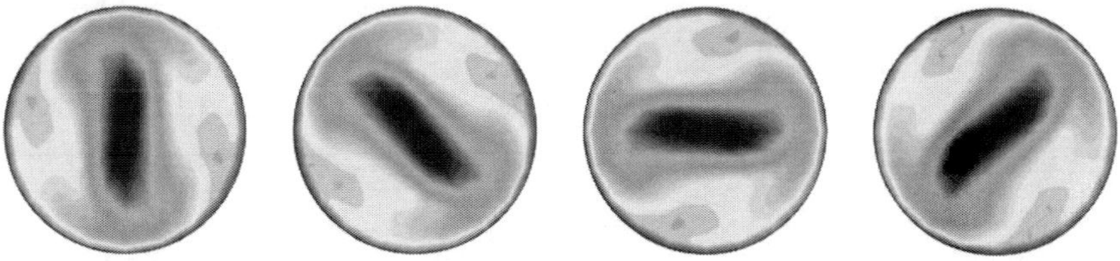

Figure 7. Snapshot of temperature distribution in a section perpendicular to the liquid bridge axis ($A_H = 0.4$, $Pr = 30$, $Ma = 3.6x10^4$, $m = 2$, Traveling Wave, red and blue colors correspond to regions of relatively hot and cold fluid, respectively).

Having completed a characterization of these dynamics from a mathematical point of view, a further understanding of the instability is now gained by considering the *mechanism* by which the two counteracting hydrothermal waves postulated for the derivation of eqs (12)-(14) can be *initiated* in the physical reality (and lead to the subsequent stages of evolution with pulsating or rotating flow features, such as those shown in Figure 4-5 and Figures 6-7 for m = 1 and m = 2, respectively).

In particular, following Chun (1980), we introduce a relatively simple physical interpretation for such mechanisms by considering a localized disturbance on the free surface of the liquid bridge. As originally argued by this author, a temperature disturbance emerging spontaneously on the surface can immediately induce a convective disturbance in the velocity field given it ability to produce locally a temperature gradient in the circumferential direction and therefore a gradient of the surface tension. The modification of the velocity field with respect to that existing prior to the spontaneous emergence of the temperature disturbance, in turn, can alter the temperature distribution. This process can enable and sustain a feedback mechanism by which the original disturbance is amplified or mitigated depending on the considered conditions.

This is shown in detail in Figure 8a-8d. An initial hot disturbance can generate two opposite surface currents moving far away from it. For continuity, liquid present in the bulk tends to move along a radial direction from the internal region (which, as explained before, is generally colder than the surface) towards the position of the initially hot disturbance. As a result, the radial current of cold flow can turn the initial hot disturbance into a cold perturbation. Once a cold spot has been produced, it can induce a modification of the axial component of the surface velocity (Figure 8e). Indeed, the strength of surface flow axially directed from the hot disk to the cold one will increase locally (in a position located upstream with respect to the disturbance). This final effect can overheat the initial cold disturbances, thereby leading the system to recover its initial condition with the hot spot located in that position (Zeng et al., 2004).

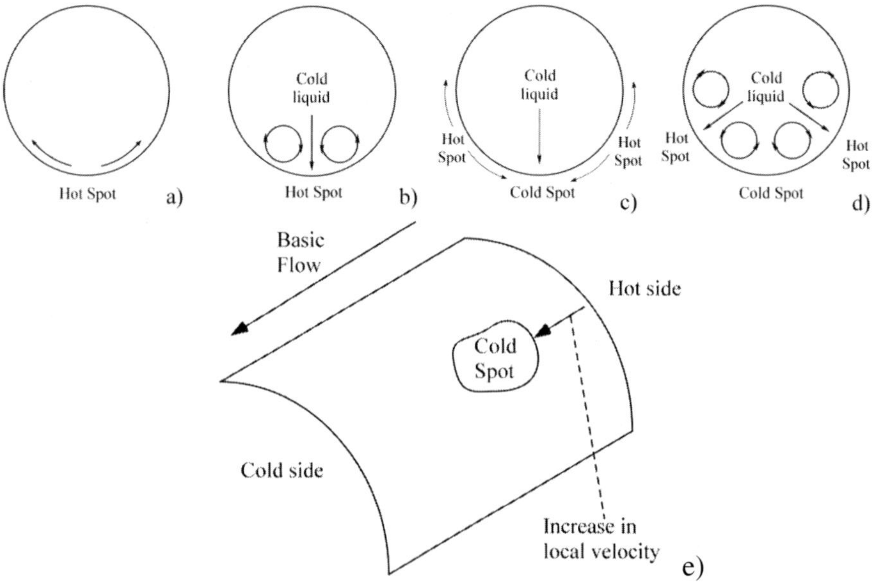

Figure 8. Amplification mechanism (sketch) leading to the emergence of counter-propagating hydrothermal waves (Marangoni convection in liquid bridges).

These dynamics are not possible for low-Pr because as in the case of liquid metals or semiconductor melts, owing to the relatively high thermal diffusivity of the liquid, any temperature disturbance would quickly be damped, thereby preventing it from coupling with convective effects as illustrated in Figure 8. By contrast, for Pr≥O(1) the mechanism summarized in Figure 8a-8e by which the hot and cold disturbances replace periodically can be turned into an everlasting phenomenon. This figure also makes evident how a disturbance initially emerging at a given azimuthal position can indirectly induce disturbances at other positions, thereby leading to the emergence of a spatially extended *pattern*, i.e., a given number of circulations in the generic section perpendicular to the liquid bridge axis (a total of $2m$ convective cells, m cells with the fluid circulating in the clockwise direction and m cells oriented in the opposite direction).

Interestingly, some useful considerations on the relationship between the azimuthal wavenumber m and the aspect ratio of the liquid bridge can be introduced on the basis of relatively simple geometrical arguments. As illustrated experimentally by Preisser et al. (1983), the radial penetration depth of convection in the bulk of the liquid is approximately given by the extension along z of the toroidal convection roll (L_V); as schematically shown in Figure 8d, the azimuthal extension of the convective cells induced by the flow instability in the cross sections perpendicular to the liquid zone axis is also approximately equal to the radial extension (L_V) of the toroidal vortex; this leads to the conclusion that the $2m$ circulations cells cover a circumference $2mL_V$. By expressing the circumference of the toroidal vortex as πD_V where D_V is its diameter, these arguments finally give rise to the following analytical relationship:

$$mA_V = \frac{\pi}{2} \rightarrow mA_H \cong 1 \tag{15}$$

where $A_V = L_V/D_V$ and $A_H = L/D$ are the aspect ratio of the Marangoni toroidal vortex and of the half zone, respectively. The agreement between this formula and available results is witnessed by both existing experiments (Figure 9) and numerical results (see, e.g., Figure 10)

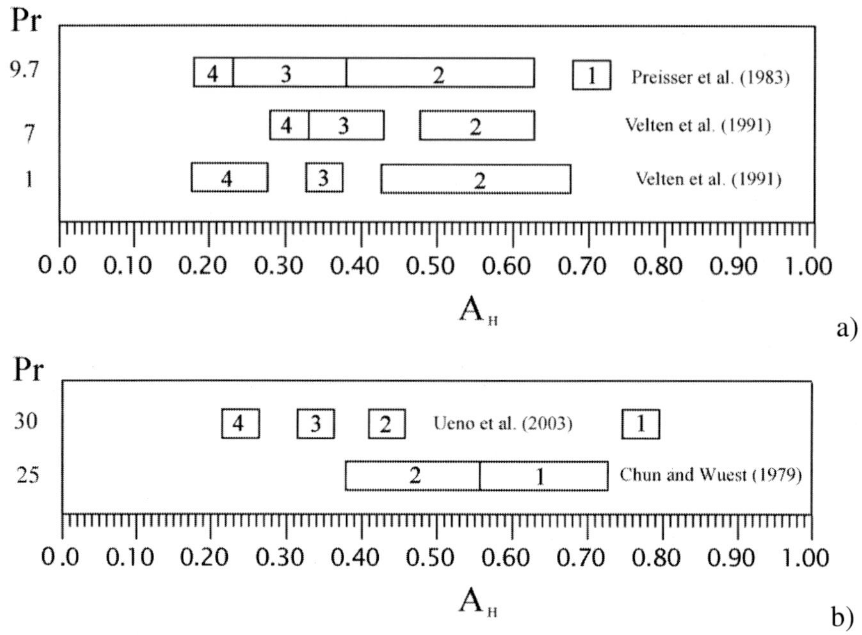

Figure 9. Critical azimuthal wavenumber versus the aspect ratio A_H for different values of the Prandtl number Pr≥1 as determined in landmark experiments (with liquid bridges heated from above in normal gravity conditions): a) Pr < 10, b) Pr > 10.

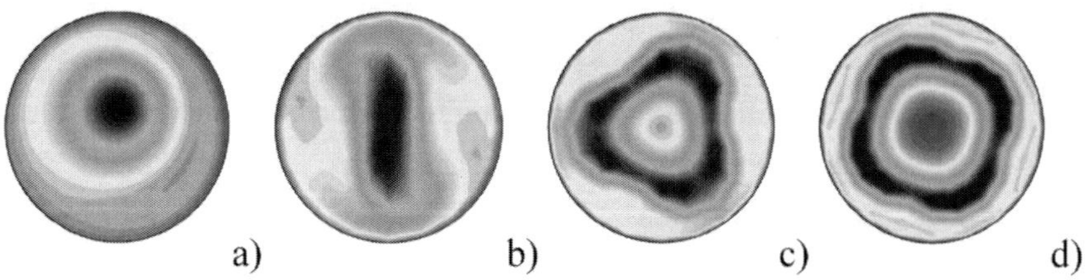

Figure 10. Snapshots of temperature distribution in a section perpendicular to the liquid bridge axis for decreasing values of the aspect ratio (Pr = 30, red and blue colors correspond to regions of relatively hot and cold fluid, respectively): a) $A_H = 1$, Ma = $3x10^4$, m = 1; b) $A_H = 0.4$, Ma = $3.6x10^4$, m = 2; c) $A_H = 0.25$, Ma = $3.8x10^4$, m = 3; d) $A_H = 0.2$, Ma = $3.9x10^4$, m = 4.

Interestingly, as shown in Figure 10, when the aspect ratio of the liquid bridge is decreased and, accordingly, the azimuthal wavenumber becomes higher, the pattern formed by the isolines of the temperature field evolves from the eccentric cold circular region visible in Figure 10a for m = 1 to a configuration displaying a regular polygonal structure (with number of sides equal to the wavenumber, as shown in Figures 10c and 10d, respectively), passing through a stage in which the shape of the inner cold region is elliptic (Figure 10b, m = 2).

For the sake of completeness (and to promote cross comparison with some of the dynamics which will be illustrated for the full zone in Sect. 4), however, we should expressly mention that the liquid bridge problem is known to gain complexity as the Marangoni number is further increased. In particular, when the Marangoni number *largely exceeds* the critical value, supercritical states with *coexisting fundamental modes* become possible (the resulting pattern in the cross section displays at the same time features relating to two or more different values of m, that is, the temperature distribution can give rise to complex polygonal irregular structures resulting from the superposition of different geometric shapes such as the fundamental ones shown in Figure 10). In order to understand this concept, it is sufficient to recall that the loss of symmetry generally implies the existence of a new solution that bifurcates from the initial state due to the selection and ensuing amplification of disturbances. Every time that the system undergoes a bifurcation, new disturbances are excited (a new mode m). If the characteristic number is increased many times, the bifurcations become faster (bifurcations sequence) and faster until the system becomes chaotic from both spatial and temporal points of view. The interested reader being referred to the studies by Frank and Schwabe (1997, 1999), Ueno et al. (2003ab), Shevtsova et al. (2003) for additional details.

3.2. Thermogravitational Flow

Though the amount of literature dedicated to the analysis of Marangoni flow in liquid bridges is really impressive, unfortunately only a handful of results have been produced for what concerns the development of convective modes of *purely gravitational nature* in this specific configuration.

Obviously, as the interface is assumed to be adiabatic for the half zone, such modes can emerge only if heating is provided from below (top disk cold, bottom disk hot), the opposite situation (top disk hot, bottom disk cold) representing an intrinsically stable state (able to maintain initially quiescent and thermally diffusive conditions ad infinitum).

As another important distinguishing mark, buoyancy convection (Rayleigh-Bénard flow) can be excited only *if a given threshold is exceeded in terms of Rayleigh number* (unlike Marangoni flow for which *no threshold needs to be exceeded to produce axisymmetric fluid motion,* Lappa, 2009). A further difference is represented by the nature of the emerging thermogravitational flow, this being always steady and three-dimensional over a wide range of aspect ratios. In the following we refer to the linear stability analysis by Wanschura et al. (1996)

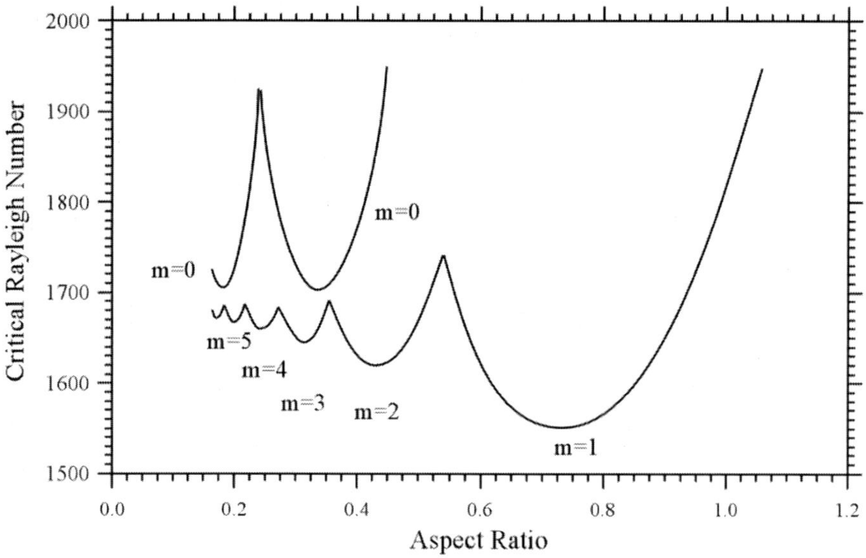

Figure 11. Stability limits for Rayleigh-Bénard convection in liquid bridges (Ra_{cr} based on the axial distance between the disks; courtesy of H. Kuhlmann).

These authors found a dependence of the critical wavenumber m on the liquid bridge aspect ratio such as that reported in Figure 11 (where it is also shown that the wavenumber m = 0, corresponding to axisymmetric flow, can enter the dynamics only for slightly larger values of Ra).

The origin of the convective disturbance pertaining to this specific mode of convection is much more intuitive if compared to the companion problem relating to the genesis of the waves in Marangoni flow. Yet, it can be illustrated in a relatively simple way by considering the impact of a localized hot disturbance on an initial quiescent fluid with unstable temperature stratification (i.e., a linear temperature distribution, see Figure 12).

Being lighter than the surrounding fluid, the parcel of fluids affected by the disturbance will obviously tend to rise, i.e., a rising current will be generated locally. As a result (due to continuity) other relatively warm fluid located underneath will be pulled up, thereby leading once again to a self-sustaining mechanism in which rising currents transporting hot fluid from the bottom towards the top cold boundary alternate in space with descending currents transporting relatively cold fluid in the opposite direction.

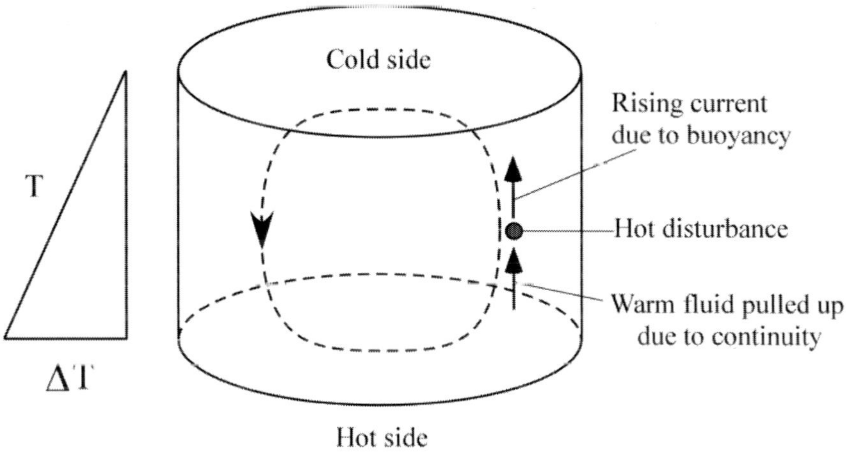

Figure 12. Amplification mechanism (sketch) leading to the emergence of Rayleigh-Bénard convection in a liquid bridge (heated from below, cooled from above and with adiabatic free surface).

The notable difference relating to the origin of the 3D flow for Marangoni and RB convection also extends to the structure of related disturbances. Additional insights along these lines can directly be gathered from Figure 5.

Figure 13. Isosurfaces of azimuthal velocity component for Rayleigh-Bénard convection in liquid bridge ($A_H = 1$, $Pr = 28.1$, $Ra = 4x10^3$, steady state with m = 1, red and blue colors correspond to regions of positive and negative azimuthal velocity, respectively).

While for Marangoni flow the isosurfaces of the azimuthal component of velocity look like curved "pillows" located in proximity to the free surface, for Rayleigh-Bénard convection they take the shape of 'bells' or multi-petal 'flowers' (with axes perpendicular to the axial direction and radial penetration depth equal to the radius of the liquid bridge, Figure 13).

Distinguishing marks, however, are not limited to the structure and stationary nature of the emerging 3D flow. What sets Rayleigh-Bénard (RB) convection apart, indeed, is its ability to produce *multiple states of convection*, that is, different solutions are allowed for the same (fixed) values of Ra and A_H. Such solutions exist in the space of parameters as independent attracting sets. However, the flow effectively emerging in a numerical simulation (or in the physical reality) essentially *depends on the specific initial conditions considered.* Using relevant language borrowed from the companion field relating to the analysis of non-linear systems, these multiple solutions can generally be considered as coexisting "attractors" in the space of phases.

Accordingly, they can be identified or studied resorting to a specific strategy of analysis based on the variation of the related "basin of attraction" (that is the set of initial conditions that leads the trajectory of the system in the space of phases to meet a specific attractor).

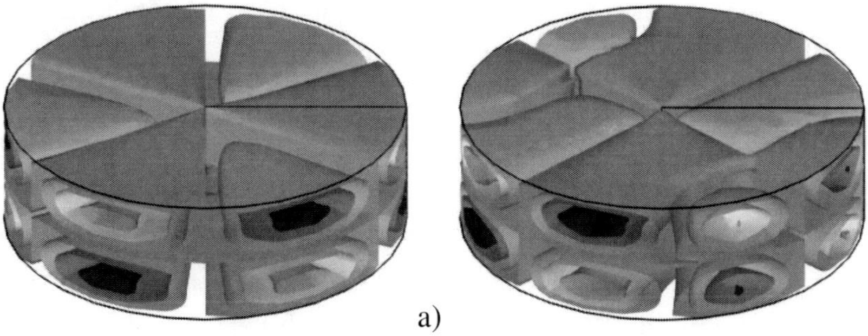

a) b)

Figure 14. Isosurfaces of azimuthal velocity component for Rayleigh-Bénard convection in liquid bridge (A_H = 0.315, Pr = 28.1, Ra = 4×10^3, red and blue colors correspond to regions of positive and negative azimuthal velocity, respectively) - Multiple steady solutions are possible: a) m = 3 (Mercedes pattern), b) four-roll state (initial conditions corresponding to 3D sinusoidal disturbances with arbitrary value of the azimuthal wavenumber added to a purely axisymmetric distribution of temperature, which increases linearly along the direction of gravity).

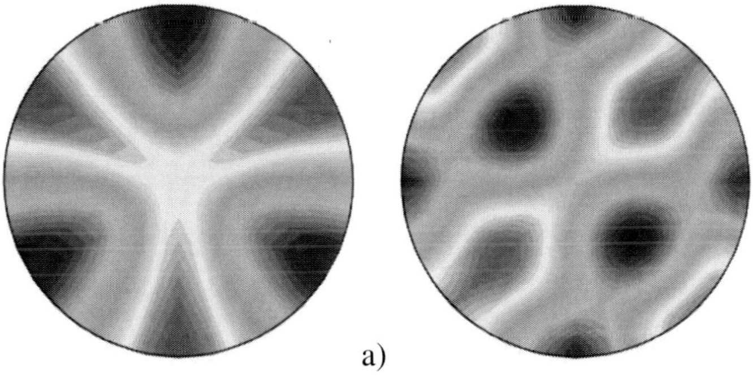

a) b)

Figure 15. Temperature distribution in a section perpendicular to the liquid bridge axis for the same conditions shown in Figure 14 (red and blue colors correspond to regions of relatively hot and cold fluid, respectively).

For instance, Figures 14 and 15 show the different equilibrium states that can be attained by a liquid bridge heated from below (Pr = 28.1) with aspect ratio A_H = 0.315, for a fixed value of the Rayleigh number (Ra = 4000) when the initial conditions are changed. While, in one case (Figures 14a and 15a) a state with three localized peaks of negative and positive azimuthal velocity can be discerned (also known as "mercedes" structure by similarity with the logo of the well-known company, Borońska and Tuckerman, 2010a,b), in the other case (Figure 14b and 15b), the central symmetry of the pattern is no longer a recognizable feature; the flow in Figure 15b, indeed consists of four rolls all approximately aligned along the same direction (flow with dominant horizontal distribution of vorticity).

For the sake of completeness, we should mention expressly that these behaviors are not an exclusive prerogative of RB convection in liquid bridges as similar studies have been produced for the case of fluids encapsulated in cylindrical containers (no free liquid-gas lateral interface).

As an example, the interested reader may consider the experimental work by Hof et al. (1999), who could observe a multitude of steady stable patterns for the same final Rayleigh number Ra = 14200 in a cylinder with A = 0.25 and a fluid with Pr = 6.7. Such states were categorized as "rolls with hot fluid rising along the center", "rolls with cold fluid falling along the center", "spoke patterns with cold fluid falling along the spokes", "spoke patterns with hot fluid rising along the spokes" and "axisymmetric pattern with hot fluid rising in the center".

Similar results have also been obtained by means of numerical simulations. As an example, Borońska and Tuckerman, 2010a,b) re-examined the case originally addressed experimentally by Hof et al. (1999), with the intent to shed some additional light on the fascinating emergence of multiple solutions reported by those authors. In the experiments by Hof et al. (1999) a large number of convective patterns had been produced by increasing and decreasing the Rayleigh number in a variety of ways. Borońska and Tuckerman (2010a) explored the same parameter space using two different types of initial conditions (namely, quiescent fluid with a linear temperature distribution in the vertical direction or fully developed flow). They initialized their numerical simulations with a slightly perturbed diffusive solution; such simulations provided different patterns, depending on Rayleigh number, which then were used as initial states *at other (smaller or larger) Rayleigh numbers*.

This study and other similar analyses have clarified that, unlike Marangoni flow in liquid bridges (which tend to favor 'axial vorticity'), RB convection emerging in cylindrical domains does not display necessarily the morphology of a toroidal roll (given its known tendency to produce parallel rolls, especially in relatively shallow containers as already shown in Figure 15b). Moreover, the liquid in the inner region can either be colder or warmer than that located more externally (for Marangoni flow in liquid bridges only the first condition is allowed).

Transition to time dependence is obviously possible also for RB convection (Lappa, 2009). For the case of convection in cylinders with aspect ratio of O(1), this transition generally occurs for relatively large values of the Rayleigh number (much higher that the value required for the onset of convection from the initial quiescent state). As illustrated by Boronska and Tuckermann (2006), in some circumstances these oscillatory modes can resemble those typical of Marangoni flow, i.e., traveling waves or standing waves.

3.3. Mixed Flow

While each of the convective effects considered in Sects. 3.1 and 3.2 on its own presents significant challenges to a full understanding of the physics controlling the behavior of the FZ, when combined they may even give rise to novel and/or unexplored phenomena. This has been clearly demonstrated by relevant studies produced over the years in which half zones supporting combined thermocapillary and thermogravitational effects (liquid bridge heated form below) have expressly been considered.

The main outcome of these studies (e.g., Wanschura et al., 1997 considered a liquid bridge with Pr = 4 and $A_H = 0.5$) is that, while in the absence of Marangoni effects there exist two equivalent convective solutions potentially emerging from the thermally diffusive (quiescent) state, which consist of flow states with up or downflow at the center of the liquid bridge (corresponding to toroidal rolls with warm or relatively cold fluid in their center, respectively), for $B_d > 1$ the presence of weak surface-tension-driven effects can break this symmetry making the bifurcation 'imperfect'.

As a result, the emergence of a single toroidal vortex with sense of circulation supported by both buoyant and thermocapillary forces (surface fluid moving from the lower hot disk towards the upper cold one) is favored (generally known as *strong state* (a), see Figure 16a).

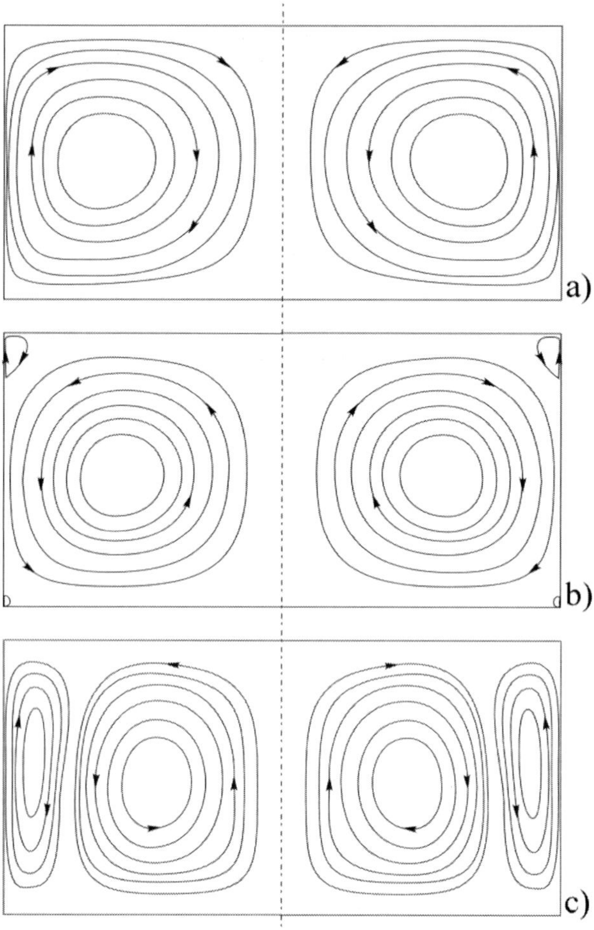

Figure 16. Multiple axisymmetric basic-flow solutions for hybrid Marangoni-buoyancy convection in half zones (for each sketch: heated and cooled disks on the bottom and on the top, respectively): (a) Strong state; (b) and (c) Weak states.

If the Rayleigh number is increased, however, other solutions become possible, namely the so-called weak states (b), and (c) shown in Figures 16b and 16c, respectively. For these states, buoyant and Marangoni effects oppose each other, and as a result, two alternate situations can be established. In one case (weak state (b)), a strong internal toroidal vortex with fluid sinking along the free surface and rising in proximity to the axis is formed, while the surface-tension driven effects are confined to a small counter-rotating cell located in the corner between the cold (top) disk and the interface. Otherwise, two separate vortices can be maintained, with an external roll (featuring fluid rising along the interface), which takes an axially stretched shape and coexists with a counter-rotating internal one.

By studying the stability of these multiple axisymmetric (basic) states with respect to 3D disturbances, Wanschura et al. (1997) found the emerging three-dimensional flows to be essentially steady. Moreover, they even identified a range of Rayleigh numbers for which the axisymmetric convection is linearly re-stabilized and the two different axisymmetric states (a) and (b) can remain stable.

Similar analyses for the opposite situation in which Marangoni effects are dominant ($B_d < 1$) have shown that, regardless of whether the liquid bridge is heated from below or from above, the set of possible 3D modes is greatly reduced from the multiplicity allowed by pure Rayleigh-Bénard flow or mixed convection with $B_d > 1$ to the only two competing solutions already discussed in Sect. 3.1 for the case of pure Marangoni convection (i.e., *standing and traveling waves*).

Rotating and pulsating regimes have been observed both experimentally and numerically for mixed thermocapillary-thermogravitational flow in half zones over relatively wide regions of the space of parameters. Among other things, these studies have led to the conclusion that pulsating or rotating waveforms are generally selected depending on the considered initial conditions (in this regard they may still be regarded as "multiple solutions" associated to a specific basin of attraction like those of RB convection) and that the former are generally unstable as they tends to be replaced by traveling waves as time increases (Lappa et al., 2000).

The two fundamental situations considered over the years for $B_d < 1$ obviously differ with regard to the aiding or opposing roles of buoyancy and thermocapillarity. For the case in which the system is heated from below, the velocity along the free surface undergoes an increase in magnitude due to the combined effect of these two forces. In the opposite conditions for which heating is applied from above, buoyancy and surface-tension-driven forces oppose each other at the free interface (in the absence of surface-tension effects the pattern would reduce to a linear thermally diffusive distribution of temperature in a motion-less liquid bridge).

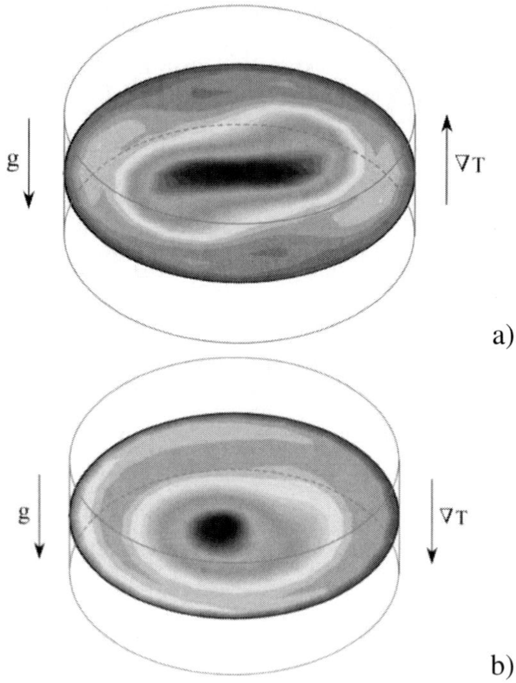

Figure 17. Snapshots of temperature distribution in a section perpendicular to the liquid bridge axis ($A_H = 0.4$, Pr = 30, liquid bridge in normal gravity conditions, red and blue colors correspond to regions of relatively hot and cold fluid, respectively): a) Ma = $3.6x10^4$, liquid heated from above, m = 2; b) Ma = $4.3x10^4$, liquid heated from below, m = 1.

In a rather counterintuitive way, mixed convection has been found to be much more stable when the heating from below is considered (Velten et al., 1991; Wanschura et al., 1997). Moreover, this condition can lead to significant changes in the emerging wavenumber with respect to the situation in which the liquid bridges is heated from above for the same set of parameters (A_H, Pr, Ra and Ma, Lappa et al., 2000, Figure 17).

Lappa et al. (2000) explained this trend on the basis of the physical mechanism responsible for the development of the instability, i.e., the sequence of stages of evolution shown in Figure 8. In particular, they ascribed the observed stabilization of Marangoni flow in liquid bridges heated from below to the expected weakened influence played by potential small surface temperature disturbances on the surface flow due to the increased strength of this flow induced by the coupling between Marangoni and buoyancy effects. An alternate point of view based on vorticity arguments was elaborated by Wanschura et al. (1997). According to these authors, the stabilization of the basic state of a half zone heated from below with respect to heating from above (or pure Marangoni flow) should be interpreted taking into account the incompatible nature of the axial vorticity associated with Marangoni flow and the horizontal vorticity favored by buoyancy convection (two mechanisms excluding each other thereby leading to stabilization of the basic state).

4. The Full Zone

As illustrated in Sect. 3, we now are in a situation where the study of the half zone has reached a sort of maturity, in the sense that numerical and experimental techniques are in position to yield useful and relevant information concerning most of the questions we may ask on the problem.

The analysis of the full zone, however, is a considerably less developed area. This model does obviously introduce an additional level of complexity into the dynamics described so far (see, e.g., the experiments by Sakurai et al., 1998 and Kudo et al., 2014). As highlighted in Sect. 2.2, it carries another degree of freedom, namely the possibility for the flow to break the symmetry with respect to the equatorial plane. In the absence of gravity (pure Marangoni flow), this formally enables two different categories of 3D disturbances, i.e., modes which are *symmetric* (preserving the reflection symmetry) or *antisymmetric* (breaking the symmetry) with respect to $z = 0$ (Figure 1b).

Unfortunately, only a few theoretical studies addressing the bifurcation to 3D flow have been appearing in the literature for Pr > 1 (Chen, and Chieh, 1995; Lan, 2003; Bouizi et al., 2007; Motegi et al., 2017).

As an example, for pure Marangoni flow Bouizi et al. (2007) investigated the transition from the initial axisymmetric state to three-dimensional (3D) flow using a Chebyshev spectral numerical method over a relatively large range of Prandtl number values ($10^{-3} \leq Pr \leq 10^{2}$). It was shown that the main conclusions related to more than three decades of studies on the companion half-zone problem could qualitatively be applied to the full zone (the flow becoming unstable against oscillatory disturbances of hydrothermal nature, which manifest as “waves”). Antisymmetric modes and standing waves were found to be dominant in this study for the case of high-Pr fluids (e.g., Pr = 20 and Pr = 100).

Though, technically speaking, a classification of the disturbances based on the distinction in symmetric and antisymmetric modes would no longer be applicable to the full zone in the presence of gravity (as buoyancy effects clearly prevent the flow from retaining reflection symmetry with respect to the equatorial plane), Motegi et al. (2017) extended this nomenclature to this case loosely re-defining these two classes of disturbances as *modes for which the temperature disturbances occurring at a fixed azimuthal station above and below the equatorial plane have the same or opposite signs.*

In particular, considering conditions similar to those originally investigated by Wanschura et al. (2007) for the classical liquid bridge (half zone, see Sect. 3.3), Motegi et al. (2017) performed a linear stability analysis for a full zone with Pr = 4 and $A_F = 1$ and increasing values of the Rayleigh number. They found antisymmetric modes to be the most critical disturbances (like the case of microgravity conditions) for relatively small values of Ra (with buoyancy essentially leading to an appreciable increase in the value of the required critical Marangoni number). The antisymmetric modes, however, were observed to be taken over by symmetric disturbances as the most critical ones for a further increase in Ra (exceeding a given threshold). Moreover, this switch in the symmetry was found together with a remarkable change in the trend displayed by the critical Marangoni number, starting to behave as a decreasing function of Ra beyond such a threshold. In this range of values of the Rayleigh number, most interestingly, these authors found the slope of the curve Ma(Ra) to become very steep, which they ascribed to *a transfer of the instability source from the toroidal roll in the upper half of the liquid bridge to the one in the lower half* induced by an increase in the buoyancy effect. According to their analysis, moreover, on further increasing Ra, symmetric hydrothermal disturbances are finally replaced by a non-oscillatory 3D mode similar to those previously found for the half-zone with dominant buoyancy.

Even though at this stage, on the basis of LSA results such as those yielded by Motegi et al. (2017), the reader might have already realized the increased level of complexity due to the interaction of Marangoni and buoyancy effects in the full zone, recent experiments have shown that reality might even be more complex.

As an example, Kudo et al. (2014) opened up a new perspective on the study of situations in which the Marangoni flow is dominant with the experimental discovery of a sort of chaotic state (for Pr = 28.1) in a region of the space of parameters where according to existing theories and earlier research studies for the half zone, mixed thermocapillary-thermo-gravitational convection should be laminar with a regular structure in space and a periodic behavior in time (single-frequency and single wavenumber flow).

Most interestingly, in such experiments, disturbances with different azimuthal wavenumbers m = 1, 2, and 3 were found *to coexist for slightly supercritical conditions* (i.e., as soon as the flow becomes unstable). Moreover, these modal structures were observed to behave in a relatively random (unpredictable) way, *switching irregularly from the typical dynamics of a standing wave to those of traveling waves and vice versa.*

These remarkable results have demonstrated that circumstances exist for which classical models might inherently be flawed.

Lappa (2016) addressed this apparent conundrum through solution of the Navier-Stokes equations (eqs. (1)-(3)) in their complete, time-dependent and non-linear form. For the same cases considered by Kudo et al. (2014), notably, these numerical results have shown that this specific problem may yet been interpreted in the frame of models and theories about the existence of 'multiple solutions'.

As explained in Sects. 3.2 and 3.3, modes of convection, existing independently in the space of states, can manifest separately depending on the initial conditions. However, if the critical conditions for their onset are relatively close, they can also be excited at the same time thereby leading to *hybrid states*.

This concept, which is not straightforward as one would imagine, can be developed in a proper way starting from the description of the specific structure of the flow established in the generic meridian plane of the full zone (Figure 18).

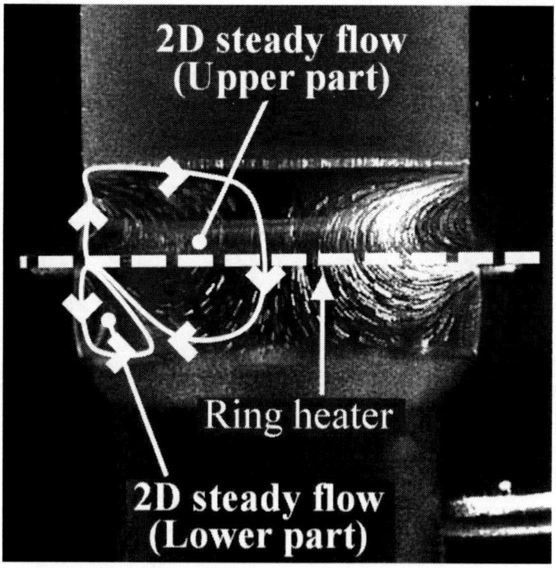

Figure 18. Typical structure of oscillatory Marangoni convection in the meridian plane of a full zone; Pr = 28.1, $A_F = 0.68$; courtesy of I. Ueno).

4.1. Spatial Flow Structure

As shown in Figure 18, in place of the single roll typically established in the half zone (Figure 3), two overlying toroidal rolls with different strength and morphological properties represent the dominant convective systems in the full zone. The different flow structure and intensity are obviously related to the different interplay of the buoyancy and surface-tension-driven effects according to the considered region.

Roughly speaking, the upper half of the full zone can be seen as a liquid bridge heated from below, and vice versa for what concerns the lower half (akin to a half zone with heating from above). Notably, reversal of the dominant temperature gradient when crossing the equatorial plane (Figure 1b) implies that buoyancy can play the role of a force *aiding* surface Marangoni flow or *counteracting* it depending on the considered half (the reader being also referred to the similar arguments elaborated in Sect. 3.3). It is as a result of such effects that the upper roll is always stronger (due to concurrent Marangoni and buoyant effects) and extends to a large extent into the lower half (Kudo et al., 2014), thereby forcing the (weaker) bottom roll into a region of relatively small radial extension located in proximity to the liquid/gas interface (where it takes an axially stretched shape, Lappa, 2016).

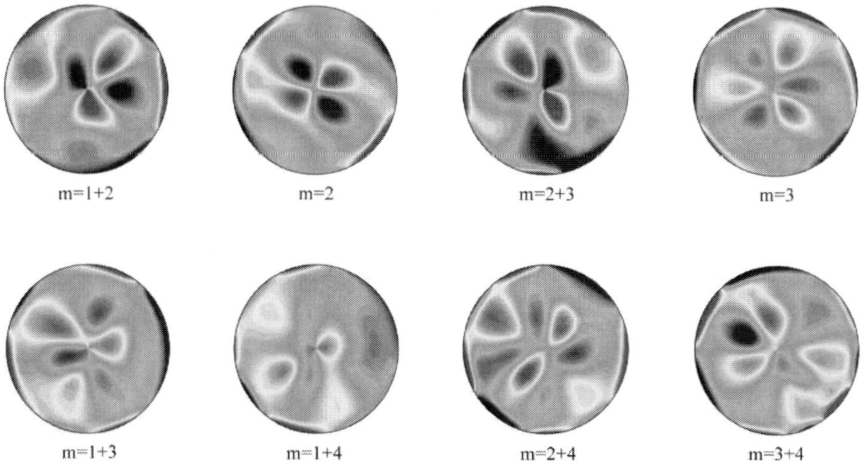

Figure 19. Snapshots of azimuthal velocity distribution at different times in cross-sections perpendicular to the full-zone axis ($z = 1/2$, upper half, $A_F = 0.63$, $Pr = 28.1$, $Ma \cong 2.3x10^4$, $Ra \cong 4.2x10^3$, red and blue colors correspond to regions of positive and negative azimuthal velocity, respectively).

In order to deal with such intricacies, Lappa (2016) examined the multicellular structures emerging due to the flow instability in selected cross sections (perpendicular to the z-axis) of the two opposing toroidal rolls located in the upper and lower halves of the full zone. For Pr = 28.1, A_F = 0.63, Ma ≅ $2.3x10^4$, Ra ≅ $4.2x10^3$, as an example, the 'multiplicity' of the involved azimuthal modes was found to be N = 4, with the azimuthal wavenumbers m = 1, 2, 3 and 4 manifesting at selected instants in isolated or combined form. Such instantaneous states are shown in Figures 19 and 20. These figures also provide interesting information on the relative amplitude of disturbances in the two opposing halves of the full zone. Indeed, by denoting with $f_{(down)m}(r,z,t)$ and $f_{(up)m}(r,z,t)$ the amplitude of the generic disturbance with azimuthal wavenumber m in the lower and upper roll, respectively, by cross-comparison of sections at z = 1/2 and z = -1/2, the reader will realize that for a given z > 0, the following inequality holds:

$$f_{(up)m}(r,z,t) \neq f_{(down)m}(r,-z,t), \quad \text{with} \quad \left|f_{(down)m}(r,-z,t)\right| \leq \left|f_{(up)m}(r,z,t)\right| \tag{16}$$

which in some circumstances and for some specific modes (compare Figures 19 and 20) even reduces to

$$f_{(down)m}(r,-z,t) \cong 0 \tag{17}$$

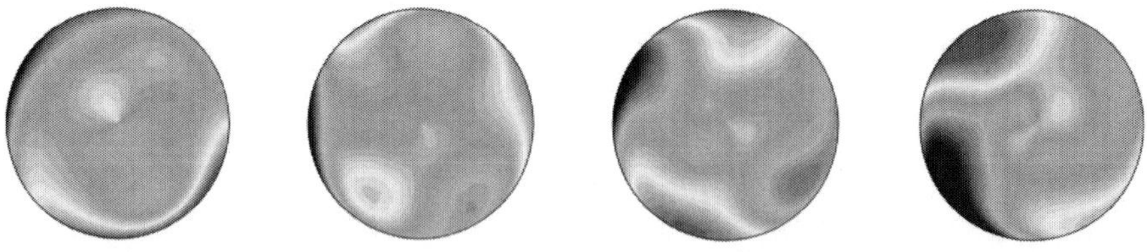

Figure 20. Snapshots of azimuthal velocity distribution at different times in cross-sections perpendicular to the full-zone axis (z = -1/2, lower half, A_F = 0.63, Pr = 28.1, Ma ≅ $2.3x10^4$, Ra ≅ $4.2x10^3$, red and blue colors correspond to regions of positive and negative azimuthal velocity, respectively).

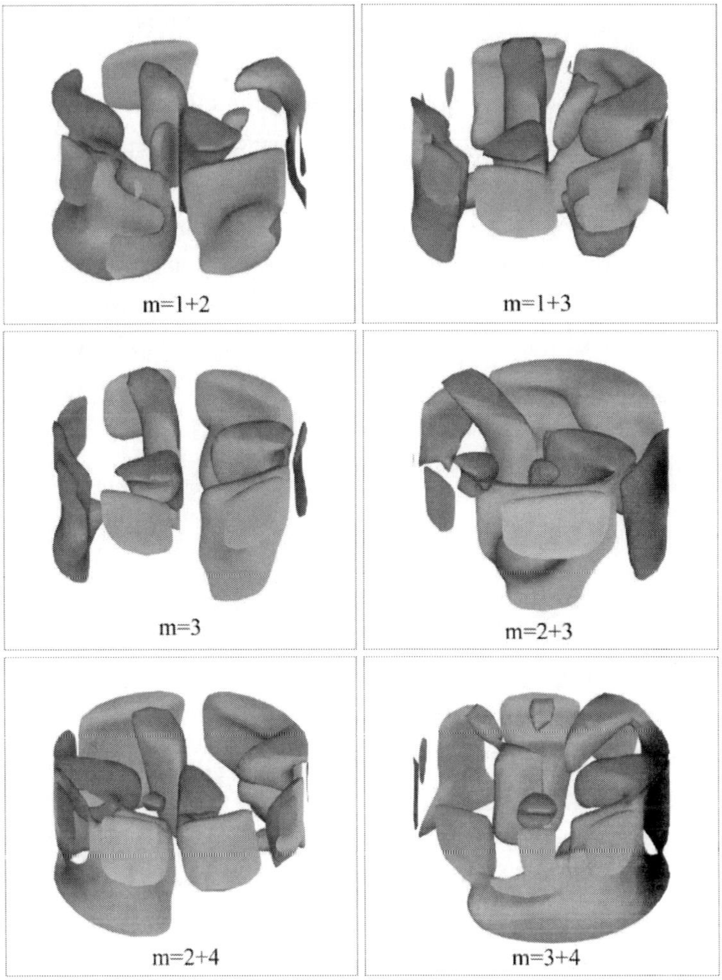

Figure 21. Isosurfaces of azimuthal velocity (snapshots) at different times ($A_F = 0.63$, $Pr = 28.1$, $Ma \cong 2.3x10^4$,$Ra \cong 4.2x10^3$). The isosurfaces correspond to three distinct values of the nondimensional azimuthal velocity (-7x10, 2x10, 1.1x10).

These results are particularly interesting for the insights they provide into the ability of coexisting toroidal rolls to increase the complexity in terms of disturbance evolution and related patterning behavior. In particular, while for the lower roll a limited number of modes is excited (disturbances with m = 1 and m = 2 in disjoint or combined form with some occasional manifestations of the m = 3 mode), a rich set of azimuthal wavenumbers is allowed for the upper roll (m = 1,2,3,4 and related combinations or 'hybrid states' as explained before). The same conclusion can be drawn via careful analysis of the isosurfaces of the azimuthal velocity in space (Figure 21), where an abrupt change in the spatial structure of disturbances can be seen when the equatorial plane is crossed.

While perturbation 'nodes' are evident for $z > 0$ in the form of disjoint 'pillows' with different colors located along the circumferential direction, an almost patternless state is established for $z < 0$. As formalized by eq. (17), no disturbances with m = 3 and m = 4 can be identified under the equatorial plane (this equation being satisfied essentially in the large wavenumber part of the spectrum of disturbances).

4.2. Temporal Dynamics

Also in this case additional useful information on the considered phenomena can be obtained by resorting to an alternate point of view based on the analysis of temperature signals. Following the same approach illustrated in Sect. 3.1 for the half zone, the history of temperature recorded at fixed positions (by experimental or 'numerical' probes) distributed along the azimuthal direction (Figure 22) can help to discern the prevailing spatio-temporal behavior. Along these lines, Figure 23 obtained for the same case shown in Figures 19-21, is particularly meaningful.

Scattering of peaks for $z = 1/2$ (Figure 23a) as opposed to ordered accumulation of minima and maxima at certain temporal locations for $z = -1/2$ (Figure 23b), indeed, indicates that 'on average' the lower roll is characterized by a pulsating behavior, whereas the upper one supports disturbances propagating circumferentially (as already explained to a certain extent in Sect. 3.1, recognizable clustering of maxima and minima should be seen as the typical signature of spatially fixed disturbance nodes, whereas significant scattering of such extrema should be regarded as evidence for traveling waves).

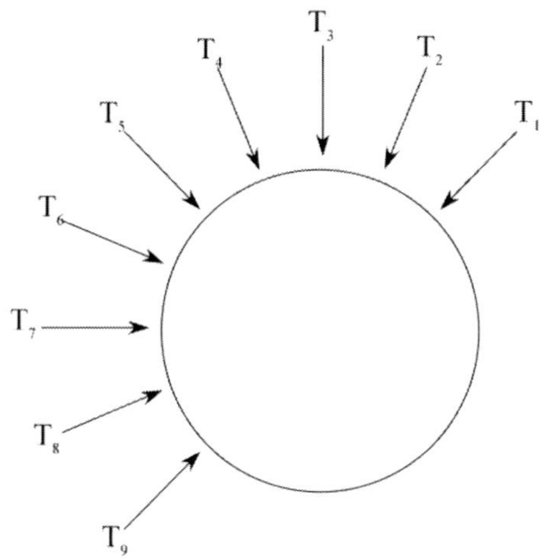

Figure 22. Sketch showing nine numerical thermocouples equally spaced along the azimuthal direction covering an angular extension of 180°.

Still following Lappa (2016), an explanation for this scenario can be elaborated in its simplest form resorting to a spatial perspective based on the simple realization that, as clarified in Sect. 4.1, the two overlying toroidal rolls have different properties.

Notably, if these rolls were completely independent, on the basis of past findings for the classical liquid bridge (Sect. 3), as a natural consequence of their different strength and geometrical properties they should undergo rather a different evolution. This argument might be used *as a first key argument to explain the observed differences in terms of prevailing disturbances in the two halves of the full zone and related spatio-temporal dynamics*.

Superimposed on these aspects, however, is the fact that the two rolls are not uncoupled, as *they can interact and exchange momentum and heat*.

In earlier studies focused on liquid metals (see, e.g., Lappa, 2003, 2004ab, 2005 for Marangoni flow in a full zone made of silicon melt), it was found that when a two-roll configuration is considered, the interaction between the two opposing toroidal convective systems can cause a significant decrease in the value of the critical Marangoni number with respect to the liquid-bridge case and make the supercritical state significantly more complex. A similar trend was also highlighted by Gelfgat et al. (2000), who examined buoyancy convection in a cylindrical configuration relevant to the so-called vertical Bridgman crystal growth method. These authors identified several azimuthal modes which become critical at relatively close values of the Rayleigh number.

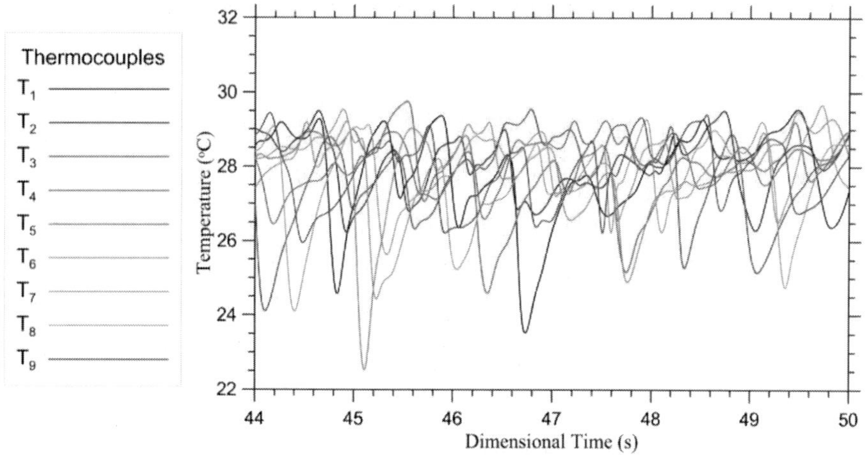

a)

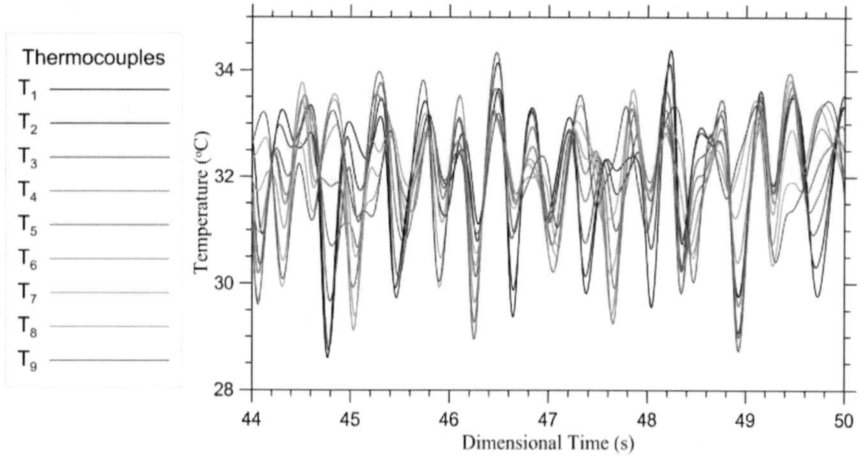

b)

Figure 23. Signals provided by the numerical probes shown in Figure 22 ($A_F = 0.63$, $Ma \cong 2.3x10^4$, $Ra \cong 4.2x10^3$): a) $z = 1/2$ (upper half), a continuous phase shift is allowed; b) $z = -1/2$ (lower half), only discrete values of phase shift can be seen.

These effects can still be invoked to justify the peculiar pattern-forming mechanism seen for the complete floating zone for Pr > 1. As explained before, two superposed toroidal rolls are embedded in the full zone, and it can be argued that their interaction can significantly expand the set of allowed instability modes (thereby lowering the critical threshold). Moreover, the multiplicity of solutions and their diversification according to the roll (upper or lower), can be justified according to the different geometrical properties of such convective systems and the effective interplay between gravitational and thermocapillary effects (which, in turn, depend on the relative position of the considered liquid region with respect to the equatorial plane).

The same arguments can also be used to explain the differences relating to the temporal behavior. According to Figure 23a, the prevailing flow in the upper larger roll can be interpreted as a combination of waves traveling with distinct azimuthal wavenumbers (m = 1,2,3,4) and different amplitudes, as witnessed by the variety of phase shifts displayed by the signals. By contrast, owing to the peculiar geometric configuration of the lower roll (axially stretched and with limited radial extension), the variety of modes is largely reduced (m = 1,2) and only two phase shifts are allowed, which indicates the coexistence of these disturbances results in standing-wave modes (waves that overlap with similar amplitude thereby leading to an essentially pulsating behavior).

As a final aspect deserving discussion, we limit ourselves to recalling that these arguments can also be used to shed some light on the apparently "intermittent" response of the system observed by Kudo et al. (2014). The random switching from the dominant rotating (pulsating) mode of convection to the pulsating (rotating) one can obviously be ascribed to the interference (non-linear interaction) among all such modes. Indeed, the related interconnected feedback loops can make the system extremely sensitive to the interaction between the upper and lower rolls. In turn, this effect can cause frequent changes in the *instantaneous* spatio-temporal mode of convection and allow intermediate (hybrid) situations in which the resulting scenario is apparently erratic (as confirmed by the numerical simulations by Lappa, 2016).

CONCLUSION

The production of oxide materials with the floating zone technique should be regarded as a relatively complex process if it is considered under the perspective relating to the convective modes which can potentially be excited in the melt.

This chapter has attempted to assemble a simple, physically intuitive and reasonably self-contained discussion of all the possible flow instabilities in liquid volumes with free cylindrical surface supporting both surface-tension driven and buoyancy convection.

These range from the classical axisymmetric mixed Marangoni-buoyancy convection and related supercritical three-dimensional regimes, which can be studied using a classical liquid bridge (half zone) held between differentially heated supporting disks, to the much more complex states enabled when heat is injected into the system through the lateral free surface (full zone).

The limited multiplicity of solutions known for pure or dominant Marangoni flow in half zones (just two alternate fundamental spatio-temporal modes of convection being allowed when the critical threshold for the onset of oscillatory convection is exceeded) is greatly expanded when pure buoyancy (Rayleigh-Bénard) convection is considered (liquid being uniformly heated from below). Remarkably, this is also a feature of mixed Marangoni-buoyancy convection in the full zone even though situations in which gravitational effects relatively weak with respect to thermocapillary ones are considered. Rather than being an intrinsic feature of the mixed convection, the multiplicity of modes allowed in this case results from the additional degree of freedom represented by symmetry rupture with respect to the equatorial plane and the coexistence of two toroidal rolls with different properties. Distinct azimuthal modes can become critical at the same time, thereby making the emerging flow apparently erratic in nature.

A general conclusion stemming from the present discussions is that the use of surrogate models of the real FZ allows more efficient exploration of wide regions of the space of parameters and, in this regard, the half zone has proven instrumental in unraveling processes that are interwoven or overshadowed in a real floating zone. In the future, however, additional studies will be required to explore the behavior of the full zone, which still hides a not-fully-understood competition of complex and diverse physical mechanisms. It is expected that the fundamental information gained from these investigations will be crucial in our ambitious goal to elaborate a general theoretical framework for the FZ process and derive a generalized set of principles to predict its evolution in many circumstances.

ACKNOWLEDGMENTS

This work has been partially supported by UK Engineering and Physical Sciences Research Council (EPSRC grant EP/R043167/1) in the framework of the JEREMI project.

REFERENCES

Balbashov A.M. and Egorov S.K., (1981), "Apparatus for growth of single crystals of oxide compounds by floating zone melting with radiation heating", *J. Cryst. Growth*, 52: 498–504.

Balbashov A.M., Tsvetkova A.A., Chervonenkis A.Y., (1975), "Imperfections in crystals of yttrium-iron garnet grown from nonstoichiometric melts", *Neorg. Mater.*, 11: 108–111.

Bazzi H., Nguyen C.T. and Galanis N., (1999), "Numerical simulation of oscillatory Marangoni convective flow inside a cylindrical liquid zone", *Int. J. Therm. Sci.*, 38(10): 863-878.

Benz K.W., (1990), "Factors controlling crystal perfections during growth under microgravity", *Proceedings 7th European Symposium on Materials and Fluid Sciences in Microgravity Oxford* (United Kingdom), 10/15 September, ESA SP 295: 59-61.

Boronska K. and Tuckermann L.S., (2006), "Standing and traveling waves in cylindrical Rayleigh–Bénard convection", *J. Fluid Mech.*, 559: 279-298.

Borońska K., Tuckerman L.S., (2010a), "Extreme multiplicity in cylindrical Rayleigh-Bénard convection: I. Time-dependence and oscillations", *Phys. Rev. E*, 81, 036320.

Borońska K., Tuckerman L.S., (2010b), "Extreme multiplicity in small aspect ratio Rayleigh-Bénard convection: II. Bifurcation diagram and symmetry classification", *Phys. Rev. E*, 81, 036321.

Bouizi O., Delcarte C., Kasperski G., (2007), "Stability study of the floating zone with respect to the Prandtl number value", *Phys. Fluids* 19, 114102.

Chen C.J. and Chieh H., (1995), "Measurement of the float-zone interface shape for lithium niobate", *J. Cryst. Growth*, 149: 87–95.

Chen G., Lizee A., Roux B., (1997), "Bifurcation analysis of the thermocapillary convection in cylindrical liquid bridges", *J. Cryst. Growth*, 180: 238-247.

Chun C.H. and Siekmann J., (1995), "Higher modes and their instabilities of the oscillating Marangoni convection in a large cylindrical liquid column", *Proceedings of the Norderney Symposium on Scientific Results of the German Spacelab Mission D2,* Editors P.R. Sahm, M.H. Keller, B. Schiewe, 235-241.

Chun C.H., (1980), "Experiments on steady and oscillatory temperature distribution in a floating zone due to the Marangoni convection", *Acta Astronautica*, 7: 479-488.

Cröll A., Kaiser Th., Schweizer M., Danilewsky A.N., Lauer S., Tegetmeier A., Benz K.W., (1998), "Floating-zone and floating-solution-zone growth of GaSb under microgravity", *J. Cryst. Growth*, 191: 365-376.

Dabkowska H.A. and Dabkowski A.B., (2010), "Crystal Growth of Oxides by Optical Floating Zone Technique", in Springer *Handbook of Crystal Growth* 367 (2010), pp. 367-391.

Dabkowska H.A. and Gaulin B.D., (2003), *Crystal Growth of Technologically Important Electronic Materials,* ed. by K. Byrappa, T.

Ochachi, M. Klapper, R. Fornari (Allied Publishers PVT, New Delhi 2003) pp. 341–354.

Frank S. and Schwabe D., (1997), "Temporal and spatial elements of thermocapillary convection in floating zones", *Exp. Fluids*, 23: 234-251.

Frank S. and Schwabe D., (1999), "Experiments on the transition to chaotic thermocapillary flow in floating zones under microgravity", *Adv. Space Res.*, 24(10): 1391-1396.

Gelfgat A.Yu., Bar-Yoseph P.Z., Solan A., (2000), "Axisymmetry breaking instabilities of Natural convection in a vertical Bridgman growth configuration", *J. Cryst. Growth*, 220: 316-325.

Higuchi M. and Kodaira K., (1992), "Effect of ZrO_2 addition on FZ growth of rutile single crystals", *J. Cryst. Growth*, 123: 495–499.

Hof B., Lucas, G.J. and Mullin T., (1999), "Flow state multiplicity in convection", *Phys. Fluids*, 11: 2815-2817.

Hurle D.T.J. (Ed.), (1994), *Handbook of crystal growth,* North-Holland, Amsterdam, 1994.

Kimura S. and Kitamura K., (1992), "Floating zone crystal growth and phase equilibria: A review", *J. Am. Ceram. Soc.*, 75(6): 1140–1146.

Kudo M., Ueno I. and Kawamura H., (2014), "Transition of thermocapillary convection in a full-zone liquid bridge", *Transactions of the JSME* (in Japanese), 80(812) (15 pages), [DOI: 10.1299/transjsme.2014tep0095].

Kuhlmann H.C. and Rath H.J., (1993a), "Hydrodynamic instabilities in cylindrical thermocapillary liquid bridges", *J. Fluid Mech.*, 247: 247-274.

Kuhlmann H.C. and Rath H.J., (1993b), "On the interpretation of phase measurements of oscillatory thermocapillary convection in liquid bridges", *Phys. Fluids A* 5 (9): 2117-2120.

Kuhlmann H.C., (1995), "Some scaling aspects of thermocapillary flows", *Microgravity Q.*, 5: 29-34.

Lan C.W., (2003), "Three-dimensional simulation of floating-zone crystal growth of oxide crystals", *J. Cryst. Growth*, 247: 597–612.

Lappa M., (1995), *Marangoni flow instabilities in liquid bridges in microgravity,* MSc Thesis, University of Naples.

Lappa M., (2003), "Three-dimensional numerical simulation of Marangoni flow instabilities in floating zones laterally heated by an equatorial ring", *Phys. Fluids*, 15(3): 776-789.

Lappa M., (2004a), "Combined effect of volume and gravity on the three-dimensional flow instability in non-cylindrical floating zones heated by an equatorial ring", *Phys. Fluids*, 16(2): 331-343.

Lappa M., (2004b), "Floating zones heated around the equatorial plane: models and simulations", *Microgravity Sci. Tech.*, XV/3: 36-51.

Lappa M., (2005), "Analysis of flow instabilities in convex and concave floating zones heated by an equatorial ring under microgravity conditions", *Computers & Fluids*, 34(6): 743-770.

Lappa M., (2007), "On the stability and prominent features of Marangoni flow in classical models of the Floating Zone Technique in normal and microgravity conditions", Chapter 5 (pp. 83-128) in *Studies on Flow Instabilities in Bulk Crystal Growth*, Editor Alexander Gelfgat, ISBN-10: 81-7895-277-7, ISBN-13: 978-8178952772, 2007, Transworld Research Network (Research Signpost).

Lappa M., (2009), *Thermal Convection: Patterns, Evolution and Stability,* John Wiley & Sons, Ltd (2009, Chichester, England).

Lappa M., (2012), *Rotating Thermal Flows in Natural and Industrial Processes,* John Wiley & Sons, Ltd (2012, Chichester, England).

Lappa M., (2016), "On the onset of multi-wave patterns in laterally heated floating zones for slightly supercritical conditions", *Phys. Fluids*, 28(12), 124105 (22 pages).

Lappa M. and Ferialdi H., (2019), *Oscillatory and Turbulent Flows of Liquid Metals in Differentially Heated Systems with Horizontal and Non-horizontal Walls*, this book.

Lappa M., Savino R. and Monti R., (2000), "Influence of buoyancy forces on Marangoni flow instabilities in liquid bridges", *Int. J. Num. Meth. Heat Fluid Flow*, 10 (7): 721-749.

Lappa M., Savino R. and Monti R., (2001), "Three-dimensional numerical simulation of Marangoni instabilities in liquid bridges: influence of geometrical aspect ratio", *Int. J. Num. Meth. Fluids*, 36 (1): 53-90.

Leypoldt J., Kuhlmann H.C., Rath H.J., (2000), "Three-dimensional numerical simulation of thermocapillary flows in cylindrical liquid bridges", *J. Fluid Mech*, 414: 285-314.

Melnikov D.E., Shevtsova V.M. and Legros J.C., (2005), "Route to aperiodicity followed by high Prandtl-number liquid bridge. 1-g case", *Acta Astronautica*, 56(6): 601-611.

Moest B., Glebovsky V.G., Brongersma H.H., Bergmans R.H., Denier van der Gon A.W., Semenov V.N., (1998), "Study of Pd single crystals grown by crucibleless zone melting", *J. Cryst. Growth*, 192(3-4): 410–416.

Monti R., (1987), "On the onset of the oscillatory regimes in Marangoni flows", *Acta Astronautica*, 15: 557-560.

Monti R., Albanese C., Carotenuto C., Castagnolo D., Ceglia E, (1995), "First Results from Onset Experiment during Spacelab Mission D-2", *Proceedings of the Norderney Symposium on Scientific Results of the German Spacelab Mission D2,* Editors P.R. Sahm, M.H. Keller, B. Schiewe, 247-258.

Monti R., Savino R., Lappa M. and Fortezza R., (1998), "Scientific and technological aspects of a sounding rocket experiment on oscillatory Marangoni flow", *Space Forum* (ISSN 1024-803X), 2(4): 293-318.

Motegi K., Kudo M., and Ueno I., (2017), "Linear stability of buoyant thermocapillary convection for a high-Prandtl number fluid in a laterally heated liquid bridge", *Phys. Fluids*, 29, 044106.

Muller G., (1988), "Convection and inhomogeneities in crystal growth from the melt", in H. C: Freyhard (Ed.), *Crystals: Growth, Properties and Applications* 12, Springer, Berlin.

Neitzel G.P., Chang K.T., Jancowski D.F., Mittelmann H.D., (1992), "Linear stability of thermocapillary convection in a model of the float-zone crystal-growth process", *Phys. Fluids* A 5: 108-114.

Ohta H. and Hosono H., (2004), "Transparent oxide optoelectronics", in *Materials Today,* ISSN: 1369 7021, Elsevier Ltd 2004, pp. 42-51.

Otani S., Tanaka T., Ishizawa Y., (1988), "Control of heat flow to feed rod in floating zone system", *J. Cryst. Growth*, 87: 175–179.

Preisser F., Schwabe D., Scharmann A., (1983), "Steady and oscillatory thermocapillary convection liquid columns with free cylindrical surface", *J. Fluid Mech.*, 126: 545-567.

Revcolevschi A. and Jegoudez J., (1997), "Growth of large high-Tc single crystals by the floating zone method: A review", *Progr. Mater. Sci.*, 42: 321–339.

Rivas D. and Vazquez–Espi C., (2001), "An analysis of lamp irradiation in ellipsoidal mirror furnaces", *J. Cryst. Growth*, 223: 433–445.

Sakurai M., Tamura A., Kinoshita A. and Hirata A., (1998), "Marangoni convection in a liquid bridge that is heated by a ring heater in a full zone model", *Journal of The Japan Society of Microgravity Application*, 15: 419-424.

Saurat M. and Revcolevschi A., (1971), "Preparation by the floating zone method of refractory oxide monocrystals, in particular gallium oxide, and study of some of their properties", *Revue Internationale des Haute Temperatures et des Refractaires* 8: 291-304.

Savino R., Monti R., Lappa M., Castagnolo D. and Fortezza R., (2001), "Preliminary results of the sounding rocket experiment on oscillatory Marangoni flows in liquid bridge" (PULSAR II Experiment), *52th Congress of the International Astronautical Federation,* Tolosa, Oct. 1-5 2001, IAF-01-J.3.03.

Schwabe D., Preisser F., Scharmann A., (1982), "Verification of the oscillatory state of thermocapillary convection in a floating zone under low gravity", *Acta Astronautica* 9 (4): 265-273.

Schwabe D. and Scharmann A., (1984), "Microgravity experiments on the transition from laminar to oscillatory thermocapillary convection in floating zones", *Adv. Space Res.*, 4(5): 43-47.

Schwabe D., (2002), "Standing waves of oscillatory thermocapillary convection in floating zones under microgravity observed in the experiment maus G141", *Adv. Space Res.*, 29(4): 651-660.

Schwabe D., (2005), "Hydrothermal waves in a liquid bridge with aspect ratio near the Rayleigh limit under microgravity", *Phys Fluids*, 17, 112104 *(8 pages).*

Shevtsova V., Mialdun A., Kawamura H., Ueno I., Nishino K. and Lappa M., (2011), "Onset of Hydrothermal Instability in Liquid Bridge. Experimental Benchmark", *Fluid Dyn. Mater. Process*, 7(1): 1-28.

Shevtsova V.M., Melnikov D.E. and Legros J.C., (2003), "Multistability of oscillatory thermocapillary convection in a liquid bridge", *Phys. Rev. E*, 68 (6), 066311 (14 pages).

Shevtsova V.M., Melnikov D.E., Legros J.C., (2001), "Three-dimensional simulations of hydrodynamical instability in liquid bridges: influence of temperature-dependent viscosity", *Phys Fluids,* 13: 2851-2865.

Shindo I., (1980), "Determination of the phase diagram by the slow cooling float zone method: The system MgO-TiO_2", *J. Cryst. Growth*, 50: 839–851.

Shindo I., Ii N., Kitamura K., Kimura S., (1979), "Single crystal growth of substituted yttrium iron garnets $Y_3Fe_5{-}x(Ga,Al)_xO_{12}$ by the floating zone method", *J. Cryst. Growth*, 46: 307–313.

Stadler A., (2012), "Transparent Conducting Oxides - An Up-To-Date Overview", *Materials*, 5: 661-683.

Tang Z.M., Hu W.R., Chen G., Roux B., (1997), "Numerical simulation of 3D thermocapillary oscillatory convection", *Microgravity Q.*, 7(2): 37-42.

Ueno I., Tanaka S. and Kawamura H., (2003a), "Oscillatory and chaotic thermocapillary convection in a half-zone liquid bridge", *Phys. Fluids*, 15(2): 408-416.

Ueno I., Tanaka S. and Kawamura H., (2003b), "Various flow patterns in thermocapillary convection in half-zone liquid bridge of high Prandtl number fluid", *Adv. Space Res.*, 32(2): 143-148.

Velten R., Schwabe D., Scharmann A., (1991), "The periodic instability of thermocapillary convection in cylindrical liquid bridges", *Phys. Fluids* A 3: 267-279.

Villora E.G., Shimamura K., Yoshikawa Y., Aoki K., Ichinose N., (2004), "Large-size β-Ga_2O_3 single crystals and wafers", *J. Cryst. Growth*, 270: 420–426.

Wanschura M., Kuhlmann H.C., Rath H.J., (1996), "Three-dimensional instability of axisymmetric buoyant convection in cylinders heated from below", *J. Fluid Mech.*, 326: 399-415.

Wanschura M., Kuhlmann H.C., Rath H.J., (1997), "Linear stability of two-dimensional combined buoyant-thermocapillary flow in cylindrical liquid bridges", *Phys. Rev. E*, 55 (6): 7036-7042.

Wanschura M., Shevtsova V., Kuhlmann H.C., Rath H.J., (1995), "Convective instability mechanism in thermocapillary liquid bridges", *Phys. Fluids*, 5: 912-925.

Watauchi S., Razzaque Sarker Md. A., Nagao M., Tanaka I., Watanabe T., Shindo I., (2012), "Crystal growth of rutile by tilting-mirror-type floating zone method", *J. Cryst. Growth*, 360(1): 105-110.

Yasuhiro S., Sato T., Imaishi N., (1997), "Three dimensional oscillatory Marangoni flow in a half zone of Pr = 1.02 fluid", *Microgravity Sci. Technol.*, X/3: 144-153.

Yasuhiro S., Imaishi N., Kuhlmann H.C., Yoda S., (1999), "Numerical simulation of three dimensional oscillatory thermocapillary flow in a half zone of Pr = 1 fluid", *Adv. Space Res.*, 24: 1385-1390.

Zeng Z., Mizuseki H., Chen J., Ichinoseki K. and Kawazoe Y., (2004), "Oscillatory Thermocapillary Convection in Liquid Bridge under Microgravity", *Materials Transactions*, 45 (5): 1522-1527.

Zeng Z., Mizuseki H., Higashino K., Kawazoe Y., (1999a), "Direct numerical simulation of oscillatory Marangoni convection in cylindrical liquid bridges", *J. Cryst. Growth*, 204: 395-404.

Zeng Z., Mizuseki H., Ichinoseki K., Higashino K., Kawazoe Y., (1999b), "Marangoni Convection in Half-Zone liquid Bridge", *Materials Transactions*, JIM, 40(11): 1331-1336.

In: Recent Studies in Materials Science
Editor: Patrick R. Lind
ISBN: 978-1-53615-270-8

Chapter 3

OSCILLATORY AND TURBULENT FLOWS OF LIQUID METALS IN DIFFERENTIALLY HEATED SYSTEMS WITH HORIZONTAL AND NON-HORIZONTAL WALLS

Marcello Lappa and Hermes Ferialdi *

Department of Mechanical and Aerospace Engineering,
University of Strathclyde, Glasgow, Scotland, UK

ABSTRACT

Non isothermal flows of liquid metals induced by buoyancy are central to many advanced technological applications in materials science, often at the cutting-edge of modern engineering. They have indeed a significant impact on the production of many materials obtained via the solidification of a melt. The quality and mechanical or electrical properties of the resulting solids and crystals are adversely affected by thermogravitational convection as it can induce defects in their atomic or molecular structure (this is the case, e.g., of typical crystal-growth techniques such as the

---

* Corresponding Author Email: Marcello.lappa@strath.ac.uk.

horizontal Bridgman (HB), the Floating zone (FZ) or the Czochralski (CZ) methods). The present chapter aims to present a focused review of landmark (past) and very recent contributions on the nature, structure and hierarchy of instabilities of this type of convection. In particular, starting from simple situations corresponding to steady and laminar flows and moving towards fully developed turbulence, we present the typical hydrodynamic and hydrothermal disturbances emerging in differentially heated liquid metals and clarify the relationship among their properties and general influential factors such as: the degree of confinement (aspect ratio), morphology (wall orientation in space) and spatial degrees of freedom (number of active dimensions) of the domain hosting the melt. Manifestations of these modes of convection (including, but not limited to, transverse waves travelling in the downstream or in the upstream direction, standing waves, modulated pulso-traveling disturbances, longitudinal waves and multi-wave patterns) are discussed in detail. More complex situations are placed in the context of existing theories on turbulence in fluids and treated using concepts, methods and tools typical of the chaotic systems analysis.

1. Introduction

Flows of liquid metals are pervasive in a number of technological applications in different fields. This is indeed the case of many manufacturing activities where the considered product passes through a liquid state before being solid. Relevant examples can obviously be found in the foundry and metallurgical sectors and in the cognate field of crystal growth from the melt.

Engineers have often to deal with such flows, given the impact they have on the considered processing techniques and the quality of resulting materials. This is especially true, when the considered materials are used for advanced applications where the perfection of the crystalline structure is a necessary pre-requisite (e.g., semiconductors and superconductors). Indeed, the properties of the flow, i.e., its structure in space and evolution in time, can deeply influence the solidification process (typically convective disturbances of various types can produce defects in the material chemical structure at the micro or macro scale in the form of "striations" or "segregations", respectively; Dupret and Van der Bogaert, 1994; Monberg, 1994).

In most cases, such flows are produced by buoyancy effects, i.e., they result from the action exerted by Earth's gravitational field on the small density differences which are induced in the liquid by existing temperature differences or non-uniformities. Though a vast literature has been produced to describe this kind of motion and related instabilities, liquid metals have their own characteristic modes of convection, which differentiate them from known behaviors in other (common) liquids such as water or other transparent melts (such as molten oxide materials or plastics, Lappa, 2019, Lappa and Ferialdi, 2018).

In such a context, this chapter will give the reader the opportunity to identify and explore a number of advanced topics in heat transfer and fluid flow, which relate to metal substances in the molten state. More specifically, starting from the review of existing literature and related historical background, in line with the general spirit of this book, we will consider the limitations of current knowledge and the new results that engineers are producing towards the end to fill existing gaps. Where appropriate, we will use computational fluid dynamics techniques and related results to support the discussions and provide evidence/exemplars for the described dynamics.

For the convenience of the reader, using a deductive approach, first the simplest theoretically possible situation will be considered, that is, thermogravitational flow developing in a horizontally unbounded layer of infinite extent. Then, more complex and realistic circumstances will be addressed in terms of shape of the containers and spatial degrees of freedom.

1.1. The Hadley Flow: Historical Background

Given the complex (often counter-intuitive) behavior displayed by these fluid-dynamic processes and their evolution in time, over the years the investigators have conceived simplified geometrical configurations to study the response of thermogravitational flow to changes in the applied temperature gradients.

Historically, the first model elaborated in such a process of abstraction is represented by the so-called *Hadley flow* (Hadley, 1735), namely, the horizontally elongated unicellular circulation induced by a horizontal temperature gradient in a fluid domain of finite extent (heated at one side and cooled at the other side). Over subsequent years, however, other researchers further simplified this configuration by concentrating on the central portion of such flow assumed not to be influenced by edge effects. This has naturally led to the introduction of the so-called 'infinite layer model', extensively employed for the identification of the typical modes of convection of the Hadley flow and related instabilities (Hart, 1972 and 1983; Gill, 1974; Laure and Roux, 1989; Kuo and Korpela, 1988; Gershuni et al., 1992).

Assuming an unbounded system and the related flow to be 'parallel' (i.e., flow with a single velocity component in the same direction of the applied temperature gradient), indeed, implies an important intrinsic advantage. It has proved very effective and convenient for the application of typical linear stability analysis (LSA) protocols, especially in the era when efficient computers for the numerical solution of the overarching non-linear equations (the so-called Navier-Stokes equations coupled with the energy balance equation) were not available yet.

LSA is the art of deducing fundamental information on the behavior (in terms of possible evolution) of certain solutions of these equations working on alternative versions of them derived through linearization. Though with some limitation (LSA cannot yield quantitative data on the effective intensity of the emerging flow), this approach has enjoyed a widespread use in the literature for its known ability to provide meaningful information on the fundamental modes of instability of several flows for which the 'basic' states were known.

For the Hadley flow, in particular, these modes can be categorized as 'transverse' and 'longitudinal'. This denomination originates from the orientation in space of the emerging rolls that, by adding new velocity components to the original basic velocity profile, break the uniformity of the initial state (the parallel flow).

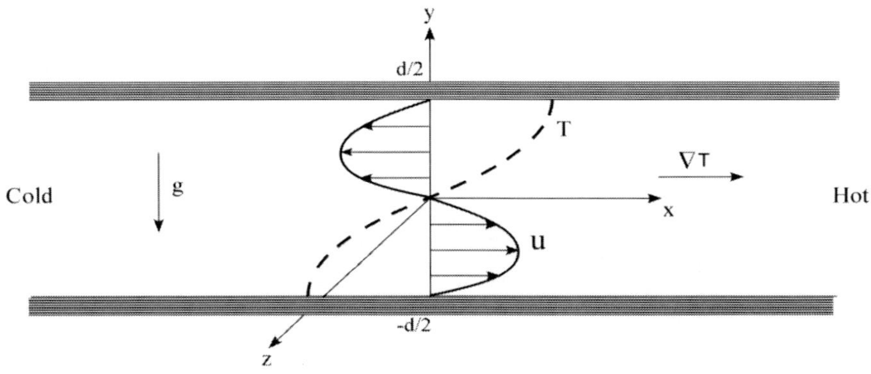

Figure 1. Basic-state velocity profile and related temperature distribution for thermal convection in an infinitely extended horizontal liquid layer delimited vertically by rigid adiabatic (thermally insulated) walls.

In other words, this means that starting from initial conditions such as those illustrated in Figure 1, the Hadley convection in an infinite layer can develop rolls with either of the orientations shown in Figure 2.

These modes differ in the spatial structure of the rolls and the angle they form with the imposed temperature gradient (transverse and longitudinal rolls having their axes perpendicular and parallel to the gradient, respectively). A significant distinguishing mark, however, also relates to the 'mechanisms' able to produce these flow features, i.e., the genesis of these convective disturbances.

The instability leading to transverse rolls is generally considered as a 'hydrodynamic' effect because it is driven by the mean shear stress that arises as a consequence of the different direction of the currents going from the hot to the cold side in proximity to the top of the layer and vice versa near the bottom. This is the reason why it is often referred to as "shear instability" and the related features as hydrodynamic disturbances. As a result, the initial 'unicellular' Hadley flow is replaced by a multicellular convective structure consisting of a chain of rolls extended along the frontier of the two opposing (primary and return) horizontal currents (Figure 2a). These rolls are stationary and become oscillatory when a second threshold in terms of applied temperature gradient is exceeded (Gelfgat et al., 1999).

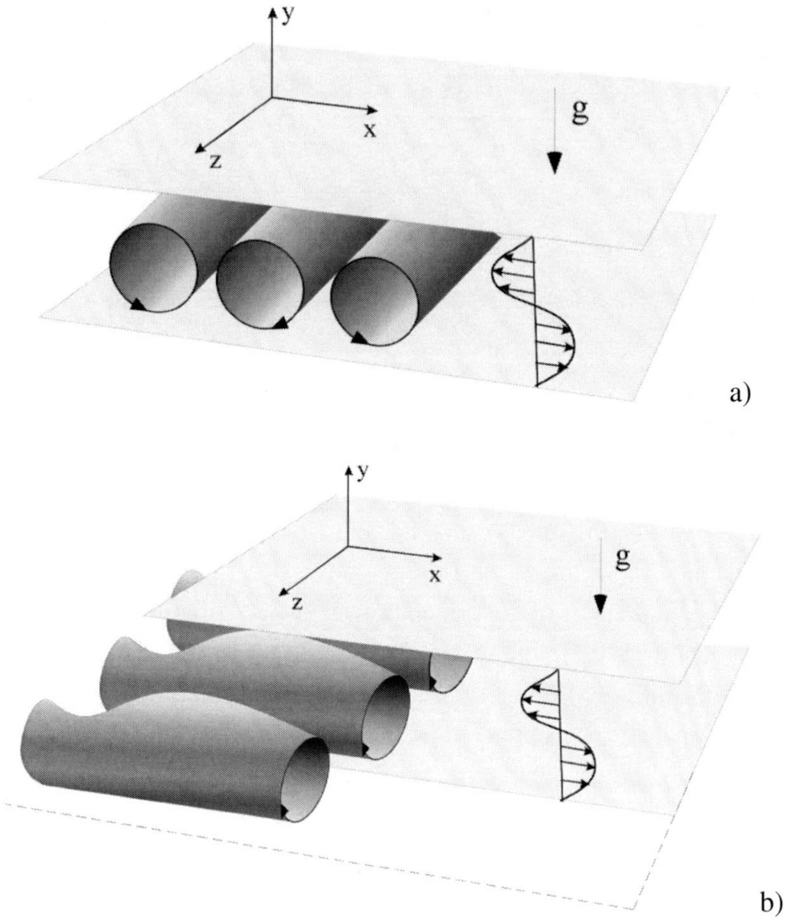

Figure 2. Sketch of convection with: a) a set of (transverse) rolls perpendicular to the imposed temperature gradient; b) a set of (longitudinal) rolls parallel to the imposed temperature gradient.

The physical process leading to the longitudinal instability is slightly more complex as it involves some interplay between the mean shear stress and the buoyancy force (a dynamical balance that makes thermal effects *directly relevant to the instability mechanism*, from which the denomination of "hydrothermal disturbances"). Hereafter, hydrodynamic and hydrothermal modes will therefore be used as synonyms for the transverse and longitudinal instabilities, respectively (the latter is also known as OLR, i.e., Oscillatory Longitudinal Roll mode of convection). Perhaps, it is also worth mentioning that in the literature the hydrothermal modes have also been called 'helical waves' given their nature of essentially travelling disturbances. This alternate nomenclature also directly follows from the superposition of the basic parallel Hadley flow and the longitudinal rolls, which force fluid particles to move along helical paths (Figure 2b).

The main results provided by the LSA for the parallel Hadley flow in an unbounded layer are summarized in Figure 3. This figure shows that different modes can exist in specific ranges of the so-called Prandtl number. This non-dimensional number, defined as the ratio of the melt kinematic viscosity and its thermal diffusivity ($Pr = \nu/\alpha$), takes a different value according to the substance considered. For liquid metals and semiconductor melts, in general, it is much smaller than 1.

In Figure 3 it can be seen that the conditions required for the onset of these convective disturbances are given in terms of critical *Grashof Number*, $Gr = g\beta_T\gamma d^4/\nu^2$, where γ is the rate of temperature increase along the horizontal direction (assumed to be constant by linear stability analyses), d is the thickness of the considered layer and β_T is the so-called thermal expansion coefficient. Sometimes, the instability thresholds are also reported in terms of the so-called Rayleigh number, related to the Grashof and Prandtl numbers by the simple relationship $Ra = GrPr$.

As evident in this figure, the emerging disturbances are essentially two-dimensional (2D) for relatively small values of the Prandtl number (transverse rolls), while they are taken over by 3D effects (OLR rolls) as Pr approaches 0.1. This observation becomes even more interesting if it is considered in combination with *existing theorems* on the stability of fluids.

The presence of inflection points in the velocity profile for purely parallel flow (see Figure 1), indeed, is known to play an important role if it is considered together with the so-called Rayleigh's theorem. According to this theorem: "*In an inviscid shear flow a necessary condition for instability is that there must be a point of inflection in the velocity profile* $u = u(y)$, *i.e., a point where* $d^2u/dy^2 = 0$". Many authors (Tollmien, 1936; Lin, 1944; Rosenbluth and Simon, 1964; Drazin and Howard, 1966) proved that this necessary condition can also be 'sufficient' in many circumstances.

Squire (1933) demonstrated that the related disturbances are purely two-dimensional.

Such arguments would indeed confirm the results of the LSA about the 2D nature of the disturbances in the limit as Pr tends to zero (inviscid flow).

The Squire's theorem, however, is no longer applicable when the Prandtl number is not exactly zero (in such a case momentum and energy cannot be uncoupled, which also invalidates the outcomes of the Rayleigh theorem). This indirectly confirms the observed 3D nature of the modes emerging for increasing values of Pr.

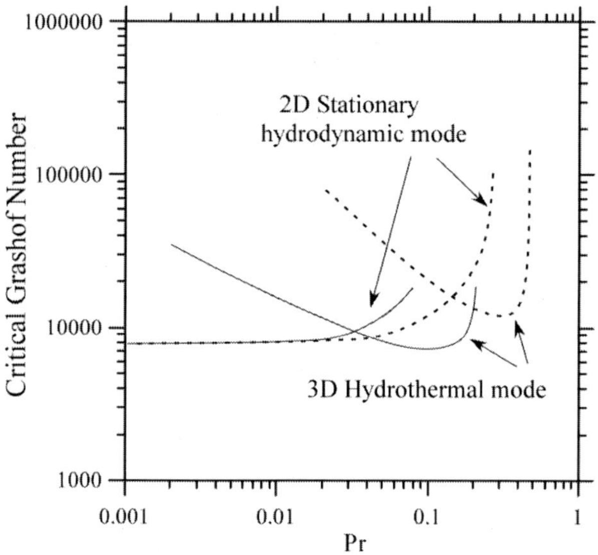

Figure 3. Stability boundaries for thermogravitational convection in an infinite layer delimited by horizontal solid walls with different thermal conditions: adiabatic (solid line) and conducting (dashed line) boundaries (Grashof number defined as Gr = $g\beta_T\gamma d^4/\nu^2$ where γ is the uniform rate of temperature increase along the x axis; courtesy of P. Laure).

Interestingly, Figure 3 also shows that the threshold values for hydrodynamic or hydrothermal disturbances depend on the nature of the thermal boundary conditions assumed for the top and bottom walls. For the case with adiabatic walls, indeed, the 3D oscillatory mode (OLR) and the 2D stationary (hydrodynamic) branches intersect at Pr = 0.034, while the intersection point is located at Pr = 0.114 when conducting walls are considered (walls with a linear profile of temperature).

Both the 2D branches (in the adiabatic and conducting cases) asymptotically tend to a unique limit (Gr = 7942) as Pr goes to zero, which provides further evidence about the origin of these disturbances, which take their energy from inertial forces of the basic flow and can therefore survive when the velocity and the thermal fields are no longer coupled (Pr = 0).

On increasing Pr from zero, the critical Grashof grows smoothly and this must be ascribed to the increasing effect of the thermal field (which, however, as explained before, plays no direct role in the hydrodynamic instability mechanism). The tide of the hydrodynamic instability branch finally becomes very steep when the Prandtl number approaches a certain value (Pr_L). This indicates that the hydrodynamic mode is no longer allowed beyond this point. This limiting value (Pr_L = 0.1 according to Kuo and Korpela, 1988 for the adiabatic case), shifts to Pr_L = 0.2 for the conducting case.

Vice versa, when the OLR branches are considered, the critical Grashof number rapidly increases when Pr tends to zero, confirming the essentially thermal nature of these modes of convection (suppressed in the limit as Pr $\rightarrow$ 0). Also for these disturbances, however, there exists a cut-off value of Pr where the curves are constrained by vertical asymptotes (located at Pr_{L2} = 0.2 for the adiabatic case, Pr_{L2} = 0.45 for the conducting case, respectively).

At this stage (for the sake of completeness), we should recall that the description of the case with conducting walls, however, would also require consideration of a third mode of convection, not reported in Figure 3. This type of convection, generally known as "Rayleigh mode" is excited by the presence of regions of unstable thermal stratification induced by the basic flow in proximity to the top and bottom wall (Gershuni et al., 1992). The related disturbances generally emerge in the form of stationary rolls with their axes parallel to the imposed temperature gradient (which explains why they are also known as SLR, i.e., Steady Longitudinal Rolls). The region of existence of these disturbances largely extends beyond the aforementioned asymptotes located at Pr_{L2} (Gershuni et al., 1992); this implies they can manifest in a variety of liquids, not necessarily liquid metals. Obviously, these states completely disappear when the horizontal walls are adiabatic (as in this case the aforementioned regions with unstable thermal stratification do not form).

1.2. Inclined Systems

For fluids with Pr < 1, the set of possible modes of convection can significantly be expanded with respect to those described in the preceding section if the considered layer *is inclined instead of being perfectly horizontal* (see Figure 4). The inclination indeed leads to the presence of a destabilizing vertical component of the imposed temperature gradient, which can induce other types of convective disturbances in liquid metals regardless of whether the top and bottom boundaries of the layer are insulated (adiabatic) or conducting.

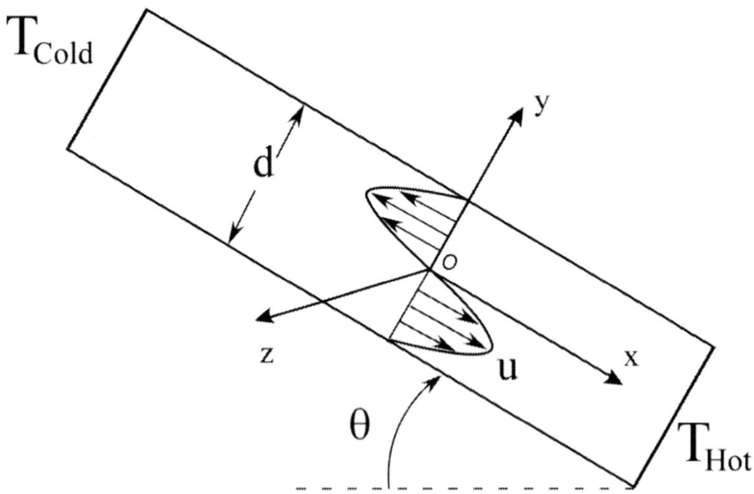

Figure 4. Sketch of differentially heated inclined rectangular box.

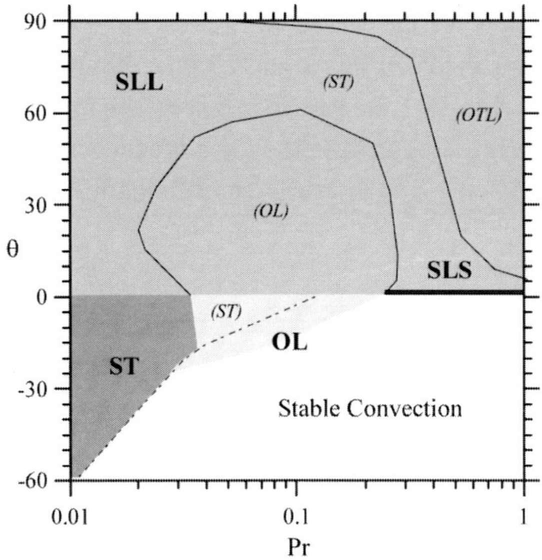

Figure 5. Instability modes in an (infinite) inclined layer with insulated walls as a function of the inclination angle and the Prandtl number (data directly provided by R. Delgado-Buscalioni). The instability with the lowest critical Rayleigh number is indicated with bold letters and that with the second lowest critical Ra is labeled in italics and between parentheses. Legend: ST (stationary transversal), OL (oscillatory longitudinal), SLL (stationary longitudinal long-wavelength), OTL (oscillatory transversal long-wavelength), SLS (stationary longitudinal short-wavelength).

The new types of instabilities in the space (θ, Pr), made possible by inclining a differentially heated cavity even if its longest walls are thermally insulated, are shown in Figure 5 and summarized in Table 1 (where some general trends have been reported for each type of instability in terms of critical Rayleigh number, defined as $Ra = g\beta_T\gamma d^4/\nu\alpha$, and wavenumber).

This problem has extensively been investigated (through LSA) by Delgado-Buscalioni and coworkers. According to their studies, the "Stationary Longitudinal long-wavelength instability" (SLL), which can appear for any $\theta < 90°$, should essentially be regarded as a 'Rayleigh mode' such as those discussed before for the horizontal layer with conducting walls. This instability is possible even if the top and bottom walls are adiabatic because by inclining the system, *the hot side can induce a heating-from-below effect.*

Table 1. Trends for the critical parameters of typical modes of instability in inclined configurations (Rayleigh number defined as $Ra = g\beta_T\gamma d^4/\nu\alpha$ where γ is the rate of temperature increase along the x axis; the configuration is assumed to have the longest walls insulated; the constants R_0, c_1, c_2 and c_3 are given in Delgado-Buscalioni, 2001)

Type of instability	Critical Ra	Critical wavenumber
ST	$Pr << 1, Ra \propto \dfrac{Pr}{\cos\theta}$ $Pr \cong 1\ (\theta < 90°),\ Ra \cong \dfrac{R_0}{\sin\theta}$	$q_x = 1.35$ $q_x = 1.6$
OTL	$Ra \cong \dfrac{R_0}{\sin\theta}$	$q_x \cong 0.3$
SLL	$Ra \cong \dfrac{Ra_{cr(\theta=0)}}{\sin\theta}$	$q_z = 0$
SLS	$Ra \propto \dfrac{Pr^{1/2}}{(c_1 - Pr)^2}$	$q_z \cong 2.9$
OL	$Ra \propto \dfrac{Pr^{1/2}}{(c_2 - Pr)^{1/2}}$	$q_z \propto \dfrac{(c_3 - Pr)^{2/5}}{Pr^{-1/2}}$

The “Oscillatory Transversal long-wavelength instability” (OTL) is another instability enabled by the inclination. It is characterized by a long wavelength (typically 9 times the height d) and can emerge only if $0° < \theta < 90°$.

The other well-known classical instabilities induced by hydrodynamic or hydrothermal (OLR) disturbances for the horizontal configuration ($\theta = 90°$, see Figure 3) are also affected by the inclination. In particular, any tilt with $\theta < 90°$ tends to expand the range of existence of the stationary transverse rolls (the 2D hydrodynamic instability, which for $\theta = 90°$, as explained before, is generally observed for fluids with $Pr < 0.034$ and is suppressed for $Pr > Pr_L \cong 0.1$). The inclination also affects the range of existence of the oscillatory longitudinal instability (OLR) (the cutoff value of Pr at which this instability is damped decreases with the inclination angle;

for example, for θ = 90° and 80° OLR perturbations are damped for Pr ≥ 0.21 and Pr ≥ 0.26, respectively).

1.3. The Non-Linear Regime: The Numerical Solution of the Navier-Stokes Equations

There is no doubt that the results provided by LSA for the horizontal or inclined infinite layer have become over the years a useful source of information for investigators interested in the fundamental gravitational modes of convection emerging in differentially heated liquid metals, and have opened new paths to a much better understanding of these fluid systems.

Unfortunately, however, the LSA is not in a condition to shed light on the behaviour of these system when they enter the non-linear regime (namely when the disturbances saturate their amplitude) as it can only provide meaningful information on the *spatial structure* of the disturbances (i.e., wavelength and wavenumber) and their *stationary or oscillatory nature*. As outlined in Sect. 1.1, estimates relating to the effective amplitude of the disturbances are out of reach because of the linearised form of the equations on which LSA relies (which prevents any information about the non-linear regime to be predicted).

This explains why, most of available results for the non-linear regime have been obtained resorting to other numerical approaches, typically based on the direct numerical solution of the Navier-Stokes equations. These equations in non-dimensional form can be cast in compact form as :

$$\underline{\nabla}\cdot\underline{V}=0 \tag{1}$$

$$\frac{\partial \underline{V}}{\partial t}=-\underline{\nabla}p-\underline{\nabla}\cdot\left[\underline{V}\underline{V}\right]+\Pr\nabla^{2}\underline{V}-\Pr Ra\; T\;\underline{i}_{g} \tag{2}$$

$$\frac{\partial T}{\partial t}+\underline{\nabla}\cdot\left[\underline{V}T\right]=\nabla^{2}T \tag{3}$$

where velocity and temperature have been referred to the scales α/d and ΔT, respectively and all distances have been scaled on d; accordingly, $\underline{V}$, T and p are the nondimensional velocity, temperature and pressure, respectively, $\underline{i}_g$ is the unit vector along the direction of gravity. Moreover, the Boussinesq approximation has been used for the buoyancy production term in the momentum equation.

For problem closure, the equations have to be complemented with relevant thermal and kinematic boundary conditions (BCs). While from a kinematic point of view, in general, these BCs simply reduce to setting $\underline{V} = 0$ on solid boundaries (the so-called no-slip property of walls), a wider set of variants is allowed for the temperature. If the considered wall is assumed to be isothermal, then a fixed constant temperature must be set along its entire extension; if it is thermally insulated, i.e., adiabatic, then the aforementioned Dirichlet condition corresponding to a fixed value of T must be turned into a Neumann condition on the value taken by the derivative of temperature with respect to a direction perpendicular to the boundary (this derivative must reduce to zero). If the wall behaves as a conducting surface, then a fixed-in-time distribution of temperature can again be assumed (typically a linear profile with the temperature increasing along the direction of the applied temperature gradient).

From a purely mathematical standpoint, the set of overarching equations (1-3) and related boundary and initial conditions constitute the so-called IBVP (Initial Boundary Value) problem, which elegantly encapsulates all the facets of this subject, namely, the role played by the presence of solid surfaces (walls), the specific three-dimensional shape (geometry) of the considered container and the initial conditions of the system.

Typical solution techniques of these equations, based on discretization of the considered physical domain, generally make possible (in a straightforward way in many cases) the consideration of domains of finite extent.

Apart from the general interest from the point of view of applied mathematics, which attaches to such a numerical problem, the IBVP allows consideration of wall-effects and geometrical constraints, which, in general, can be used to explain the observed discrepancies between the predictions of the LSA for the infinite layer and the effective behavior of finite-size systems. It can also be directly linked to another very interesting aspect concerning the properties of spatially confined systems, that is, the *existence of multiple solutions* (we will come back to this very interesting concept later in this chapter).

Though many different numerical approaches are possible, in the following we concentrate on the so-called category of *projection methods*. At this stage we should clearly point out, indeed, that the practical numerical implementation of the IBVP is not independent of the specific numerical procedure considered. This apparently counter-intuitive concept originates from the realization that the adoption of a specific solution technique enriches the IBVP with additional boundary conditions. These are generally known as "numerical boundary conditions" (NBC) to distinguish them from those dictated by the physics of the considered problem. In such a context, the choice of the numerical method becomes a delicate aspect as relevant information on the proper selection of relevant NBCs making the problem well posed form a mathematical point of view is known for some cases only.

This is why for problems dealing with the study of fluid-dynamic bifurcations and eventually the existence of "multiple solutions" (we will clarify the meaning of this concept later) we recommend the use of methods pertaining to the aforementioned set of projection techniques. All the numerical examples presented in this chapter have been obtained in the framework of such approach. For this class of methods the problem relating to NBCs making the problem well posed has already been investigated to a certain extent leading to widely recognized theoretical results (Gresho and Sani, 1987; Gresho, 1991). With these methods, moreover, the problem can be tackled directly in terms of 'primitive variables' (i.e., velocity and pressure).

These techniques may be regarded as a spin-off of the strategy originally conceived by Harlow and Welch (1965), Chorin (1968) and Temam (1968) on the basis of the well-known theorem of inverse calculus (see, e.g., Ladyzhenskaya, 1969), by which any vector function can always be decomposed into a divergence-free part $\underline{V}$ and the gradient of a scalar potential -P (a curl-free part). They are based on a simple concept, that is, the momentum equation can physically retain the correct amount of vorticity relating to the considered problem even if the contribution represented by the pressure gradient is not taken into account (from a mathematical standpoint, this property stems from the irrotational nature of every vector field resulting from the gradient of a scalar quantity):

$$\underline{V}^* = \underline{V}^n + \Delta t\left[-\underline{\nabla}\cdot\left(\underline{V}\underline{V}\right) + \Pr \ \nabla^2 \underline{V}\right]^n + \Pr Ra T^n \hat{\underline{n}} \tag{4a}$$

The integration of the truncated momentum equation obviously leads to an intermediate velocity $\underline{V}^*$, which has no physical meaning. It does not even satisfy the continuity equation, in other words, it is not solenoidal. In order to correct it and make the resulting field incompressible, projection methods introduce a new velocity field as a linear combination of the unphysical velocity field $\underline{V}^*$ and the pressure gradient previously neglected.

$$\underline{V}^{n+1} = \underline{V}^* - \Delta t \underline{\nabla} p^n \tag{4b}$$

This approach formally leads to an additional equation required to determine the pressure, which would otherwise remain unknown. This equation is formally obtained by forcing the 'corrected' velocity field to be divergence-free. This leads to a purely elliptic equation to be solved on the basis of an iterative process (see, e.g., Lappa, 1997):

$$\nabla^2 p^n = \frac{1}{\Delta t}\underline{\nabla}\cdot\underline{V}^* \tag{4c}$$

To make the problem well-posed, the initial velocity field $\underline{V}_0 = \underline{V}(t = 0)$ must correspond to a quiescent state or satisfy the condition $\underline{\nabla}\cdot\underline{V}_0 = 0$.

Obviously, the three equations given above refer to an implementation of such a method using schemes explicit in time (the superscript n there indicates the time step). However, different variants of such approach are possible according to the philosophy used to integrate the simplified momentum equation with respect to time (via explicit or implicit schemes) and the staggered or non-staggered arrangement of primitive variables on the computational grid (see, e.g., Moukalled et al., 2016). Moreover, the accuracy of such a method obviously depends on the nature and order of the schemes used to integrate effectively the above equations and the related time and space integrations steps (such aspects are not covered here not to increase excessively the length of the chapter; the interested reader being referred to the relevant literature for additional details, e.g., Lappa and Ferialdi, 2017 and 2018b).

2. Spatially Confined Systems

2.1. Geometrical Effects and Multiplicity of Solutions

Owing to the availability of methods for the solution of the governing equations in their complete (non-linear and time-dependent) form such as those illustrated in Sect. 1.3, over the years, configurations much more realistic than the infinite layer have been considered. Related studies have led to the more or less widespread realization that the ranges of Prandtl number and Grashof (or Raylcigh) number in which the hydrodynamic or hydrothermal modes of convection can be found depend on the effective geometry of the container hosting the melt.

An archetypal numerical study along these lines is due to Afrid and Zebib (1990), who could clearly show that the effective extension of the physical domain in the spanwise direction (the direction perpendicular to the basic flow, i.e., the z axis in Figure 1) determines a significant variation in the threshold value of the Rayleigh number. The main outcomes of these numerical simulations were confirmed by effective experiments with opaque liquid metals (see, e.g., Hung and Andereck (1988) and Pratte and Hart (1990) for $Pr = 0.026$).

All these studies, conducted considering classical rectangular cavities with vertical (heated or cooled) and horizontal walls, provided reasonable hints that the location of the points in the (Ra, Pr) space where the branches pertaining to the different instability mechanisms (transverse or longitudinal) intersect might not an invariant property.

In such a context, some authors expressly focused on the possible interplay of these two fundamental modes of convection. As an example, Braunsfurth and Mullin (1996) could identify experimentally four 'apparently new' modes of oscillation for liquid metals with $0.016 \leq Pr \leq 0.022$. Similarly, through parametric numerical simulations in the interval $0 \leq Pr \leq 0.027$, Wakitani (2001) found the value of the critical Rayleigh number to grow on increasing Pr and/or by reducing the extension of the cavity in the spanwise direction. Most interestingly, these variations were *not regular*, which was interpreted as evidence for the co-existence of different oscillatory modes competing at the onset. The presence of coexisting modes of convection was also reported for supercritical conditions in the 3D numerical studies of Delgado Buscalioni et al. (2001) for inclined cavities.

Leaving aside for a while these interesting results, it should be mentioned that another interesting line of inquiry has been originated by the realisation that all these influential factors [namely, the considered melt (Pr), the control parameter (Gr or Ra), the effective spatial configuration of the container (physical domain geometry), nature of boundary conditions (top and bottom walls with adiabatic or conducting conditions) and eventually the inclination of the overall system (θ)] may not be the only ones able to affect the solutions of the aforementioned well-posed (Lappa, 2002) IBVP (Initial Boundary Value).

The initial boundary conditions can also play a significant role.

This has been revealed by a set of numerical studies originally conducted under the assumption of 2D flow (Crespo del Arco et al., 1989; Pulicani et al., 1990; Okada and Ozoe, 1993ab and Gelfgat et al., 1999), where hydrodynamic disturbances (the only ones allowed in those circumstances) were found to produce a variety of potential behaviors, including multi-roll states, the bifurcation from steady to time-periodic solutions, the emergence of hybrid modes of convection due to the co-existence and interaction of multiple states and backward transitions from multi-roll to unicellular regimes. The most interesting of such features was, perhaps, the "multiplicity" of possible solutions, that is, the existence of distinct branches of solutions *for a fixed set of parameters.*

These studies clarified that, though such branches can exist as independent sets of solutions, the flow effectively emerging in a numerical simulation (or in the physical reality) essentially *depends on the specific initial conditions considered.*

In other words, even if the fluid (Pr), the geometry (cavity aspect ratio), the boundary conditions, the applied Gr (or Ra) and possible disturbances (hydrodynamic modes only for 2D systems) are *fixed*, the emerging flow can change in morphology (number of rolls) and nature (stationary or oscillatory) depending on the initial state of the system. Using relevant language borrowed from the companion field relating to the analysis of non-linear systems, these multiple solutions can generally be considered as coexisting 'attractors' in the space of phases (that is a space having as many dimensions as the degree of freedom of the considered system). Yet using specific technical language, it is known that in a problem developing coexisting branches of solutions, the trajectory of the system in the space of phases can tend selectively to a different attractor according to the considered "basin of attraction", which, by definition, is *the set of initial conditions that leads the system trajectory to meet a specific attractor.* An example of such behavior is shown in Figure 6.

If the numerical simulation is started from thermally diffusive conditions ($\underline{V}_0 = 0$), for a relatively low Rayleigh number, i.e., Ra = 10^3, the flow is simply given by a 'twisted' recirculation embracing three co-rotating (anticlockwise in the figures) rolls; this kind of flow is still a solution if, starting from the conditions shown in Figure 6a ($\underline{V}_0 \neq 0$), the Rayleigh number is increased to 2000. However, if the same simulation for Ra = 2000 is repeated assuming initial diffusive conditions, a double-vortex configuration can also be obtained; these flows are both steady and *centrally symmetric* (namely, the pattern does not change with respect to rotation through 180° about the center of the cavity).

This simple example illustrates why in order to identify the window in which the system is subjected to multiple coexisting attractors, a typical strategy of analysis must be based on the exploration of the system response to a variation of the initial conditions. Put in practice, in addition to 'standard' initial conditions corresponding to a quiescent state with a diffusive distribution of temperature, other possible initial conditions must be considered (typically the solutions corresponding to one branch can be used as the initial conditions for the determination of the solutions relating to another branch, see, e.g., Kengne et al. (2018), Gelfgat et al. (1999).

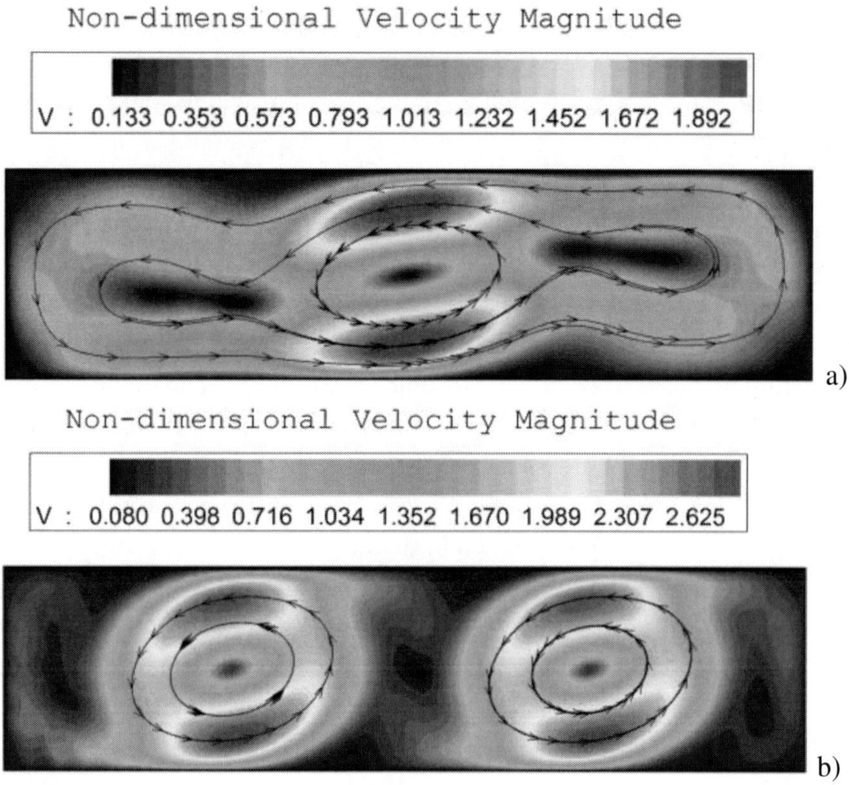

Figure 6. Multiple states of two-dimensional steady convection originating from initially thermally diffusive and quiescent conditions for a rectangular cavity with aspect ratio $A_x = L/d = 4$ (cold and hot sides on the left and on the right of each frame, respectively; upper and lower boundaries with adiabatic conditions, Rayleigh number defined as $Ra = g\beta_T \Delta T d^3/\nu\alpha$ where ΔT is the horizontal temperature difference): a) Single elongated roll for $Ra = 1x10^3$, b) two-roll state for $Ra = 2x10^3$.

2.2. Systems with Converging or Diverging Walls

This section builds on, but also seeks to extend, the valuable research that has highlighted the importance of the spatial confinement on the properties of the thermogravitational flow of liquid metals in differentially heated enclosures. In the remainder of this chapter, in particular, we concentrate on the very recent findings by Lappa and Ferialdi (2017 and 2018b) about configurations with *oppositely inclined walls* (Figure 7).

These systems may be regarded as an additional variant with respect to those traditionally considered in the past. What sets them apart is that, though the top and bottom walls are not horizontal, *no net inclination is present* (as these boundaries are one the mirror image of the other). Related studies pertain to a new line of inquiry in which, though the system is not tilted, the flow cannot be considered 'parallel'.

The geometrical configuration under examination (shown in Figure 7) is a shallow cavity symmetric with respect to the xz plane with converging or diverging walls. Its average depth is denoted by d. The lateral solid walls perpendicular to the x-axis are assumed to maintain a fixed temperature (one side cooled, the other side heated, having height d_{hot} and d_{cold}, respectively).

Three different non-dimensional (independent) parameters must therefore be introduced to characterize this problem, namely the streamwise (or transverse) aspect ratio A_x, defined as the cavity length-to-average-depth ratio $A_x = L/d$ where $d = (d_{hot} + d_{cold})/2$, the spanwise (or longitudinal) aspect ratio, $A_z = W/d$ (where W is the extension of the computational domain along z) and what we call the expansion (compression) ratio $\eta = d_{hot}/d_{cold}$ which can be > 1 (diverging geometry) or < 1 (converging geometry). In this way a more general problem is obtained in which the classical cavity with horizontal top and bottom walls considered in earlier 2D or 3D studies is just one realization (it being recovered for $\eta = 1$).

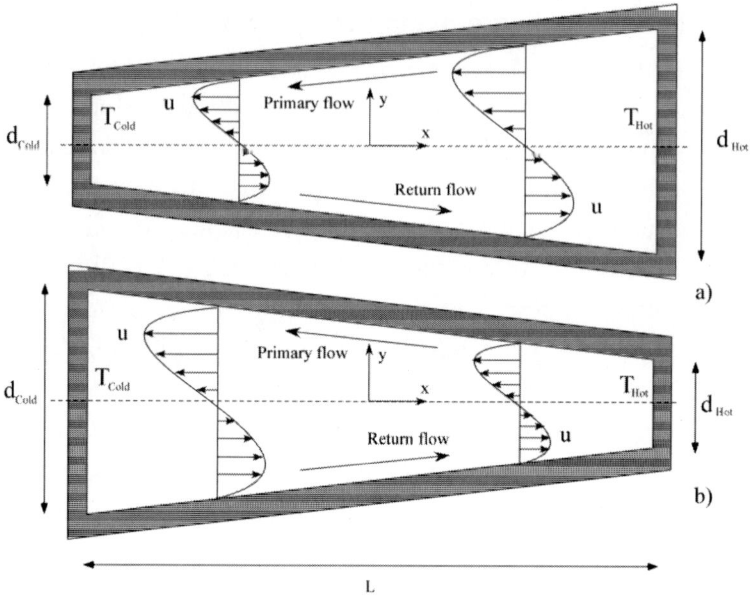

Figure 7. Sketch of enclosures with converging or diverging walls, related thermal and kinematic boundary conditions and Cartesian coordinate system: a) $\eta < 1$, b) $\eta > 1$.

In order to filter out the presence of Rayleigh modes such as those identified by Gershuni et al. (1992) for the case of conducting boundaries (which would further increase the complexity of the analysis as discussed at the end of Sect. 1.1) we assume the top and bottom walls to be adiabatic (no heat exchange). The considered system is also prevented from developing Rayleigh modes such as those reported by Delgado Buscalioni et al. (2001) for inclined cavities (see Sect. 1.2) as in the present case the heated and cooled walls remain always perfectly vertical (thereby excluding the presence of destabilizing vertical components of the temperature gradient, i.e., components potentially leading to *a heating-from-below effect*).

As sketched in Figure 7, the natural buoyancy flow emerging is such configurations for relatively small values of the Rayleigh number ($Ra = GrPr = g\beta_T\Delta Td^3/\nu\alpha$ where ΔT is the horizontal temperature difference) simply consists of a single horizontally elongated unicellular flow (with fluid moving upward in proximity to the heated wall, flowing along the upper wall from the right side to the left side, sinking when it meets the left (cold) wall and finally returning to its original position along the bottom).

Given the complexity of the matter, we are forced to articulate the presentation and discussion of the results in two main stages. More precisely, first we concentrate on 2D findings as a first necessary step to clarify the general problem relating to the existence of multiple solutions and transition to time dependence. Then the treatment is extended to fully 3D configurations.

2.2.1. The Zoo of 2D Modes

In this section, entirely devoted to the discussion of 2D results, two specific values of the aspect ratio A_x are considered for a representative small value of the Prandtl number (Pr = 0.01), namely $A_x = 4$ (already extensively investigated in past studies on the subject due to its relevance to typical industrial techniques for crystal growth) and $A_x = 10$. Moreover, the other parameter of interest, i.e., the compression (or expansion) ratio is varied in a relatively wide range ($0.1 \leq \eta \leq 10$) and the system response is described with respect to the Rayleigh number spanning two orders of magnitude, i.e., $O(10^3) \leq Ra \leq O(10^4)$.

The different results pertaining to this set of parameters for the case $A_x = 4$ can be seen in Figure 8, where the solutions have been reported as a function of η and arranged accordingly as 'columns', while the 'rows' correspond to variations in the value of Ra. By taking a look at this map, it becomes evident that, though the flow is forced to remain two-dimensional, a variety of solutions are allowed. These can be categorized as: (i) unicellular steady states (US), (ii) multi-roll steady states (MS), (iii) quasi-stationary states with weakly oscillating rolls (QS), (iv) 'classical' oscillatory rolls (with slender 'appendages' undergoing a kind of rhythmic displacement) (P), (v) states with periodically touching or 'kissing' rolls (K), (vi) regimes with periodically coalescing rolls (C), (vii) purely travelling waves (TW), (viii) mixed states (PLT) with travelling waves 'locally' modulated by roll pulsations or splitting behaviors (Lappa and Ferialdi, 2017).

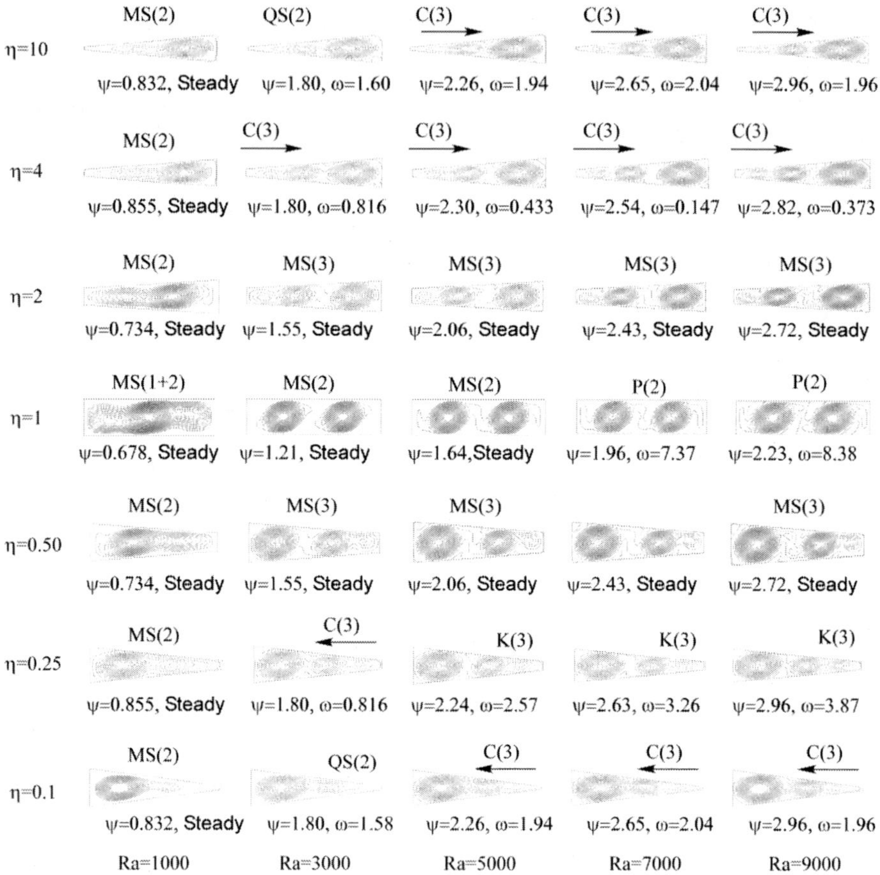

Figure 8. Map of solutions versus the parameter η and the Rayleigh number ($Ra = g\beta_T\Delta Td^3/\nu\alpha$) for $A_x = 4$ (ψ is the maximum of the streamfunction, multiple states existing for a fixed value of η have been reported as separate "rows" at the related value of η and its reverse).

While unicellular states obviously correspond to the trivial solutions for which the flow behaves as the classical circulation with opposing direct and return flows theorized by Hadley (1735), the other solutions relate to unstable states produced by the onset and development of hydrodynamic disturbances (as explained before, hydrothermal waves are obviously suppressed in this case, given the absence of the third dimension).

As expected, the emerging flow is steady for relatively small values of the Rayleigh number and it becomes oscillatory as Ra is increased (as witnessed by the value of the angular frequency ω reported next to the maximum value of the streamfunction for each case).

Prior to embarking into an exhaustive description of these oscillatory states, in the following first we provide some initial and fundamental information on the *spatial symmetry properties* displayed by the flow (this strategy will prove very useful later when we will discuss the general features of such solutions from a temporal point of view and focus on the emergence of hybrid states and related dynamics).

By carefully inspecting Figure 8, the reader will realize that the behaviors observed for a given η (e.g., $\eta = 2$) are the "mirror image" of those seen for its inverse value ($\eta = 0.5$). The number of rolls and their morphological properties are essentially the same. As an example, the rolls structures obtained for $\eta = 2$ could formally be turned into those emerging for the same value of the Rayleigh number and $\eta = 0.5$ by simply rotating the overall system by 180°. By swapping the cold side and the hot side, indeed, such a rotation would transform the initial problem with $\eta > 1$ and gravity directed downwards (where the fluid raises along the sidewall of larger size, namely the hot wall located on the right of the figure) into a completely equivalent problem for $\eta < 1$ in which gravity is directed upwards and the fluid moves in the same sense (upwards) in proximity to the cold side (the cold side being located on the right side after the 180° rotation). In Lappa and Ferialdi (2017), this property of the velocity field for $\eta \neq 1$ was named the "*180°-rotation invariance principle after η reversal*". From a mathematical standpoint such a property may be formalized as follows (the tilde being used to indicate the transformed quantities):

$$\begin{bmatrix} \tilde{u} \\ \tilde{v} \end{bmatrix}_{\eta} = \underline{\underline{R}}(180°)\begin{bmatrix} u \\ v \end{bmatrix}_{1/\eta} \tag{5a}$$

with the rotation matrix defined as

$$\underline{\underline{R}}(\theta) = \begin{bmatrix} \cos\theta & -\sin\theta \\ -\sin\theta & \cos\theta \end{bmatrix} \tag{5b}$$

In particular, the cases η = 1 might be seen as special conditions for which the "*180°-rotation invariance principle after η reversal*" would reduce to the "*central symmetry*" already discussed with regard to the examples shown in Figure 7.

Having provided a general characterization of the flow from a spatial perspective, we now turn to describing in detail the related temporal dynamics. The typical oscillatory behavior for the special condition η = 1 can be seen in Figure 9 for Ra = 9000. In agreement with earlier numerical studies about rectangular cavities, flow oscillations manifest as morphological variations periodically affecting the shape of the peripheral streamlines of each roll ((P)-mode).

As shown in Figure 8, for $A_x = 4$, this kind of oscillatory flow is still possible if values of η in a given neighbourhood of η = 1 are considered (0.75 ≤ η ≤ 1.33). In particular, for η = 0.75 (η = 1.33) and η = 0.65 (η ≅ 1.5) oscillations of the P type are excited for Ra ≥ 7000 and Ra ≥ 9000, respectively.

Moreover, though the system experiences a strong stabilization in the ranges 0.4 ≤ η < 0.75 and 1.33 < η ≤ 2.5, oscillatory behaviors are again possible for both η < 0.4 or η > 2.5. Interestingly, Lappa and Ferialdi (2017) identified transition to time dependence in these ranges of η for relatively small values of Ra (hereafter, for simplicity we will discuss the behaviors for η < 1, as those for η > 1 can be easily gathered by applying the aforementioned invariance principle).

In the intervals (0.2 ≤ η ≤ 0.3) and for Ra ≥ 3000, in particular, through systematic variation of the initial conditions and/or the use of different schemes for the integration of the overarching equations, Lappa and Ferialdi (2017) identified *two different kinds of solutions coexisting in the space of parameters*, namely states pertaining to the (K) or (C) regime (relevant examples being shown in Figures 10 and 11, respectively).

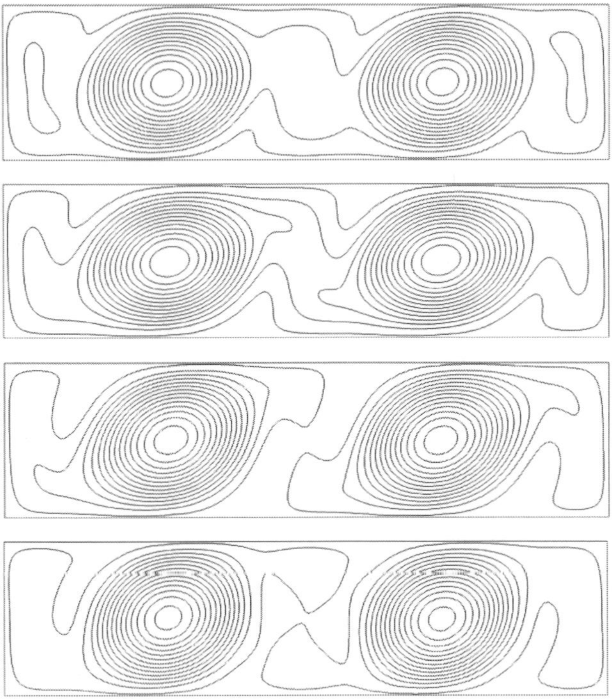

Figure 9. Oscillatory flow of the (P) type for $A_x = 4$, $\eta = 1$ and Ra = 9000: four snapshots equally spaced within one period of oscillation.

In the first case, two of the tree rolls existing at the same time in the cavity (more precisely, the central roll and the roll near the shortest wall) display a weakly oscillatory process in which one of them contacts periodically its neighbor ("kissing" rolls behavior, which provides a justification for the "K" letter to label it) whereas the third roll remains steadily located in proximity to the left (cold) wall as a separate entity.

The kissing behavior, however, is replaced by a complete roll coalescence process as solutions pertaining to the (C) branch are considered (Figure 11). The roll merging mechanism takes place essentially in the left part of the cavity as a result of the displacement of the roll initially located in the center of the cavity towards the longest vertical wall. As soon as the central and left rolls become a single entity, a new roll emerges at the right side thereby preserving the initial number of rolls and giving the observer the feeling of a *disturbance that spreads continuously from the right side to the left side* (i.e., in the direction of growing cross-sectional area).

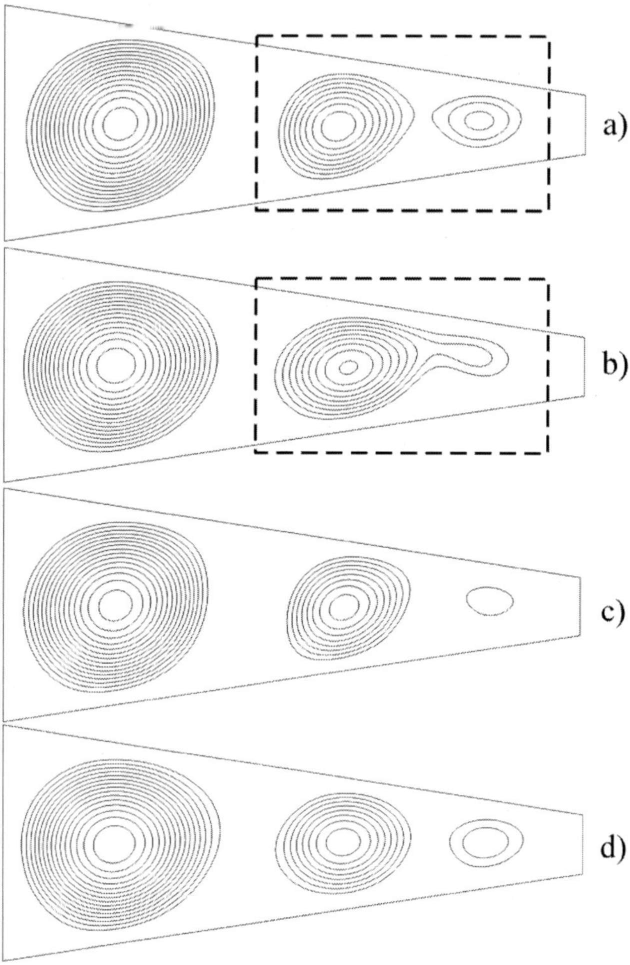

Figure 10. Oscillatory flow of the (K) type for $A_x = 4$, $\eta = 0.25$ and Ra = 9000: four snapshots equally spaced within one period of oscillation.

Lappa and Ferialdi (2017) found this alternate mode of convection, featured by the propagation of a disturbance from regions of reduced cross-sectional area towards regions where the cavity is thicker, to be *the preferred mode of oscillatory instability* in the end regions of the considered interval of η (namely when η is very small or very large). Notably, Lappa and Ferialdi (2017) also highlighted that such a travelling disturbance or wave (TW) can become the dominant mechanism of oscillatory convection through the entire range of Rayleigh numbers and values of $\eta \neq 1$ if cavities with larger aspect ratio A_x are considered.

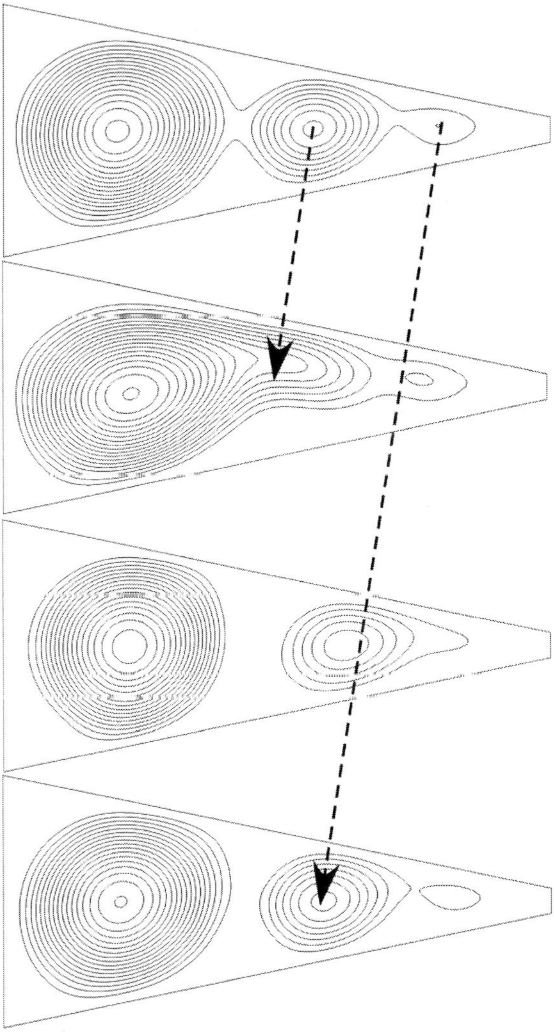

Figure 11. Oscillatory flow of the (C) type for $A_x = 4$, $\eta = 0.1$ and Ra = 9000: four snapshots equally spaced within one period of oscillation.

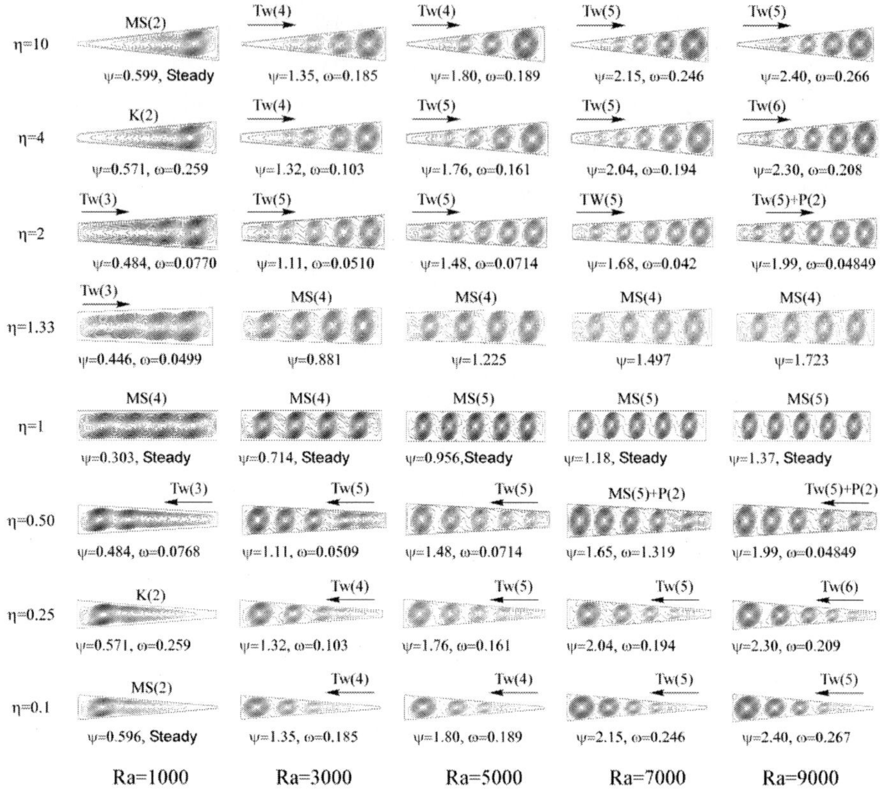

Figure 12. Map of solutions versus the parameter η and the Rayleigh number ($Ra = g\beta_T \Delta T d^3/\nu\alpha$) for $A_x = 10$ (ψ is the maximum of the streamfunction, multiple states existing for a fixed value of η have been reported as separate "columns" at the related value of η and its reverse).

Related results, summarized in Figure 12 for $A_x = 10$ using the same mode of representation already considered for Figure 8, indeed, provide evidence for such a general scenario (together with some interesting information about 'variants' and mixed states, which emerge for specific couples (η, Ra)).

As an example, for $A_x = 10$, $\eta = 0.1$ or $\eta = 10$, the initially steady unicellular flow obtained for Ra = 1000 and 2000 is quickly replaced on increasing Ra by an oscillatory mode of convection with the disturbance always clearly travelling in the direction in which the cross-section of the domain increases, namely from the hot wall to the cold wall for converging walls and in the *opposite direction* in the case of diverging walls.

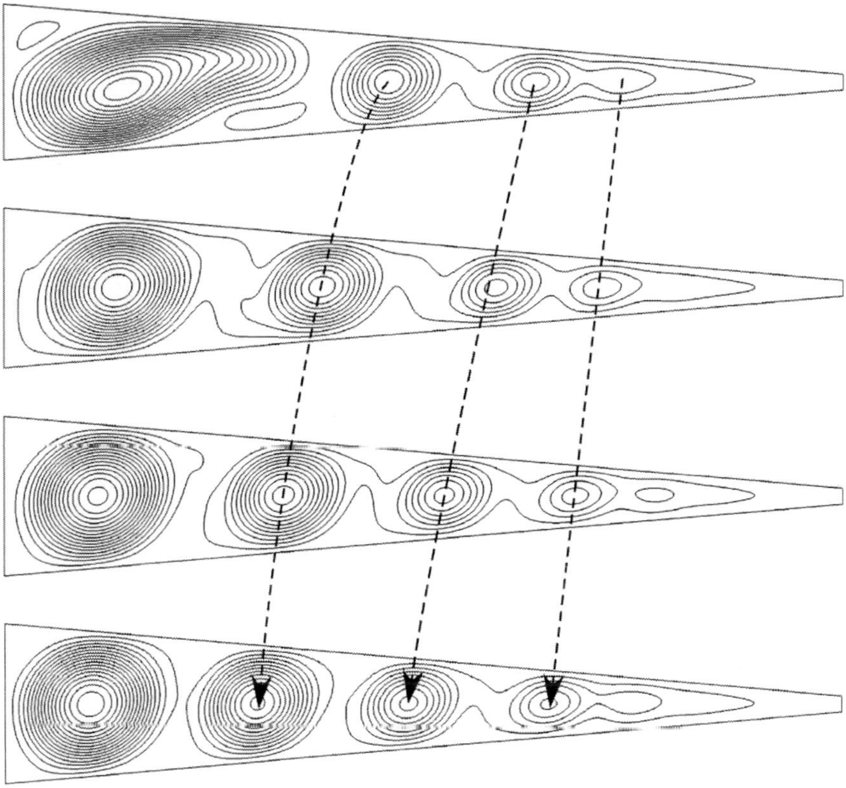

Figure 13. Oscillatory flow for $A_x = 10$, $\eta = 0.1$ and Ra = 9000: four snapshots equally spaced within one period of oscillation.

As shown in Figure 13, the oscillatory disturbance manifests itself as a well-defined chain of cells moving from the thinnest side towards the thickest one, with the exception of an almost motionless roll steadily located in proximity to the target wall, regardless of whether $\eta < 1$ or $\eta > 1$ (the stable behavior of such a stationary roll being supported by the relatively strong buoyancy forces being present in the thicker part of the domain).

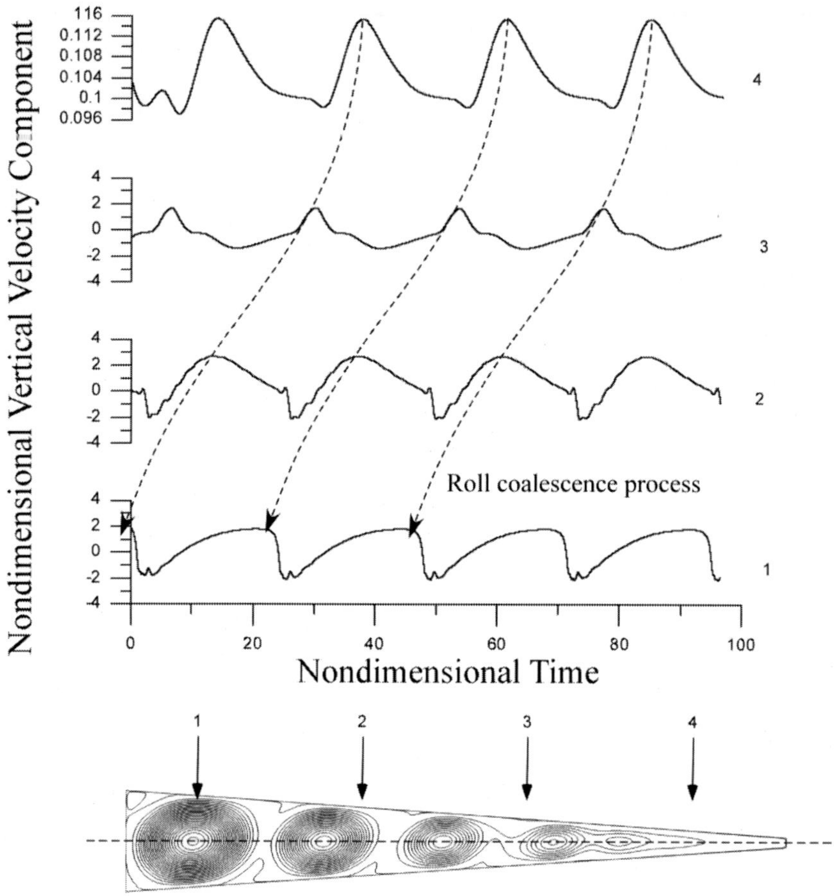

Figure 14. Velocity signals as measured by four probes periodically positioned along the horizontal direction ($A_x = 10$, $\eta = 0.1$ and Ra = 9000, vertical velocity at y = 0). The modulation of the sinusoidal signals being related to the main frequency is due to the presence of a second frequency for this relatively high value of the Rayleigh number.

As soon as one of the travelling rolls meets this roll, they merge. As already explained for the cavity with $A_x = 4$, when such a coalescence process is completed, a new roll emerges near the other sidewall, thereby *keeping constant the average number m of convective cells affecting the cavity at any instant* (and preserving the time periodicity of the overall process).

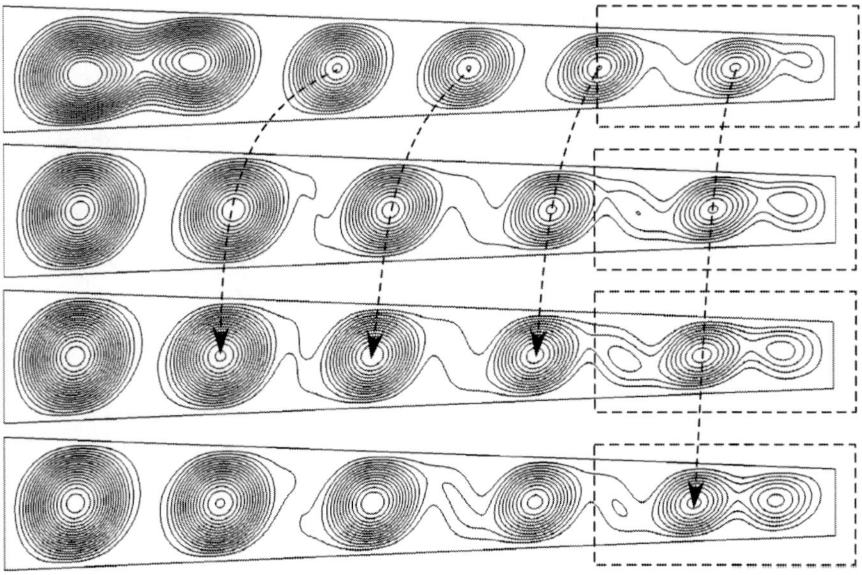

Figure 15. Oscillatory flow for $A_x = 10$, $\eta = 0.5$ and Ra = 9000: four snapshots equally spaced within one period of oscillation.

Interestingly, the velocity of propagation of the rolls is not constant along the x axis, as witnessed by the signals provided by a set of probes evenly spaced along this axis (Figure 14). Apart from confirming the travelling nature of the disturbance, made evident by the continuous phase shift displayed by the signals measured at different locations, this plot clearly shows that a moving roll experiences a varying velocity in the course of its migration.

For $A_x = 10$, Lappa and Ferialdi (2017) found the scenario for $\eta = 0.25$ and $\eta = 4$ to be very similar to that for $\eta = 0.1$ and $\eta = 10$, respectively. Nevertheless, some interesting dynamics were spotted for $\eta = 0.5$ ($\eta = 2$). According to their findings, indeed, for these values of η and relatively high values of Ra (Ra≥7000), a localized disturbance can manifest in the form of a modulation affecting the thinnest part of the cavity (highlighted by the dashed rectangle in Figure 15).

The localized oscillatory behavior is essentially due to the time-periodic formation of a three-roll structure in the right part of the physical domain (Figure 15). Apparently, this structure (resembling the 'twisted' convective structure encapsulating the three co-rotating vortices shown in Figure 6a for the purely rectangular cavity with $A_x = 4$ and $Ra = 10^3$) derives from the 'splitting' of the original roll located near the right wall (this is why this state was referred to by Lappa and Ferialdi (2017) as an m = "5+2" mode). For Ra = 9000, interestingly, this localized perturbation can exist in parallel with the travelling wave, thereby giving rise to a regime with two coexisting frequencies (one relatively small connected to the wave and another much larger one accounting for the localized disturbance of the (P) type).

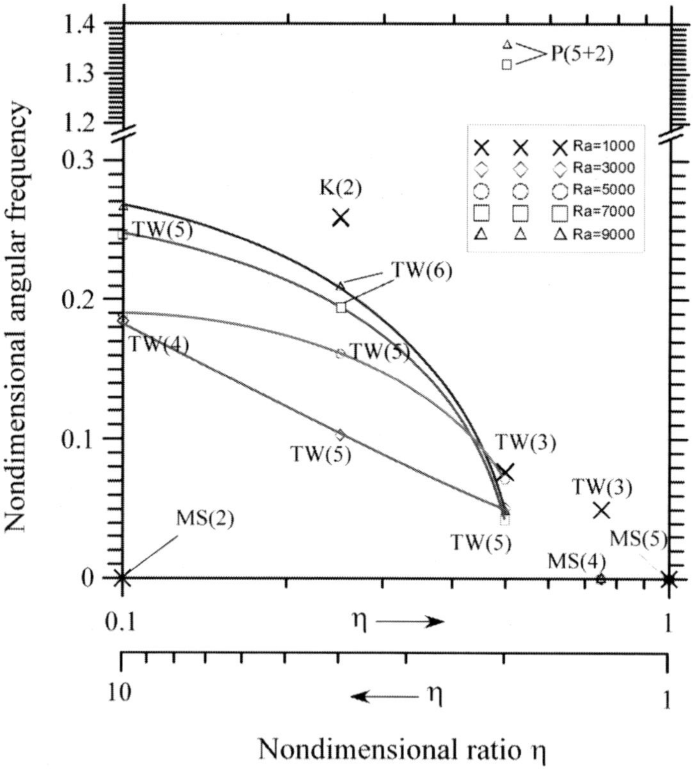

Figure 16. Oscillation angular frequency versus the expansion or compression rate η for $A_x = 10$ and different values of the Rayleigh number (for η = 1, transition to oscillatory flow occurs for Ra = 14500 with angular frequency ω ≅ 6.48, not shown in this figure).

If η is made ≅ 1, all these waveforms and related modulations effects are finally replaced by a completely different time-dependent scenario where travelling disturbances are no longer a feature of the flow (the spatio-temporal behavior becoming similar to that reported in Figure 9 for $A_x = 4$, $\eta = 1$ and Ra = 9000).

One may therefore state that, taken together, Figures 12 and 16 indicate that TW branches are dominant in the 'external regions' of the η-interval ($\eta \leq 0.5$ and $\eta \geq 2$); and that, an increase in the Rayleigh number generally causes a shrinkage of the effective η intervals in which (TW) disturbances are allowed. The latter observation is in line with the scenario already described for $A_x = 4$ (see Figure 8), that is, the preferential mechanism of oscillations is gradually transferred from low-frequency waves to high-frequency disturbances of the (P) type when Ra is increased (Lappa and Ferialdi, 2017).

2.2.2. The Zoo of 3D Modes

Having finished a description of the 2D flow in enclosures with converging or diverging walls, we turn now to considering the role played by the third dimension, i.e., the dynamics in systems spatially extended along the spanwise direction z. These cases clearly display an increased degree of complexity with respect to 2D solutions as hydrodynamic modes can no longer be an exclusive feature of the emerging flow. Shear-driven modes can indeed be completely replaced by hydrothermal modes or still be present and interact with them, leading to a variety of possible scenarios depending on the effective circumstances considered.

As in the presence of complex dynamics, in order to make the discussion simpler and clearer, a reduction of the considered ranges of parameters is beneficial, in this section we will limit ourselves to considering configurations with $A_x = 10$.

Let us recall that, as illustrated at the beginning of this chapter (Sects. 1.1 and 1.2), in past linear stability analyses based on the assumption of horizontal parallel flows, hydrodynamic and hydrothermal disturbances were observed as the preferred perturbations for $Pr = O(10^{-2})$ and $Pr = O(10^{-1})$, respectively (see Figure 3). More precisely, Hart (1983) found the hydrodynamic and hydrothermal disturbances to be the most dangerous for Pr < 0.015 and 0.015 < Pr < 0.27, respectively. Kuo and Korpela (1988) slightly revised these ranges as Pr < 0.033 and 0.033 < Pr < 0.2 (the critical curves for these two different categories of disturbances were also found to be relatively close one to another in terms of values of the Rayleigh number, the reader being referred once again to Figure 3). The influence of the system inclination on the (codimension-two) point, where the branches pertaining to the different disturbances intersect, was investigated by Delgado-Buscalioni (2001), who confirmed the existence of this point at Pr ≅ 0.03 for purely horizontal cavities, showing that it moves towards smaller values of Pr as soon as the considered configuration undergoes a *net* tilt.

As the last part of this chapter is explicitly devoted to the examination of these fundamental modes of convection in 3D geometries, we therefore focus on two representative values of the Prandtl number located in the ranges reported above, i.e., Pr = 0.01 and Pr = 0.05. In particular, to facilitate direct comparison with the results obtained under the constraint of two-dimensionality, first we describe the emerging dynamics for the same value of the Prandtl number (i.e., Pr = 0.01) already examined in Sect. 2.2.1.

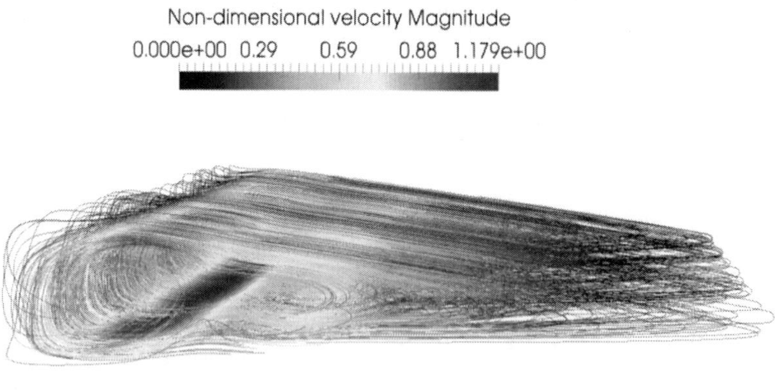

Figure 17. Steady base flow in different sections along the z direction (A_x = 10, η = 0.1, A_z = 10, periodic boundary conditions along z, Pr = 0.01, Ra = 1000, Gr =Ra/Pr= 10^5).

As a first example, Figure 17 shows the 3D field emerging for Ra = 1000 assuming $A_z = 10$, $\eta = 0.1$ and periodic boundary conditions along the z axis.

In agreement with the 2D findings (Figure 12), it can be seen that for this value of the Rayleigh number, a second circulation nucleates in the original unicellular structure and the overall flow is steady.

When Ra is increased, however, interesting departures from 2D behaviors occur.

A first evidence of such a departure is immediately provided by Figure 18, where we have reported the velocity field for Ra = 2000. By careful inspection of these plots, indeed, the reader will realize that interesting flow features along the spanwise direction z can be detected. These consist of four peaks of velocity (red arrows) located in proximity to the thickest side of the cavity. Evidence for the essentially 3D nature of the flow is also reinforced by the presence of recognizable vortices spatially extended throughout the xz plane. Moreover, this flow is essentially oscillatory.

The difference with respect to the 2D flow reported in Figure 12 becomes even more evident if one considers that, on the basis of the study conducted under the constraint of two-dimensionality, the solution should still be steady for $A_x = 10$, $\eta = 0.1$ and Ra = 2000. According to the 2D study, time-dependence should enter the dynamics at Ra $\cong$ 3000 in the form of a wave (TW) with four rolls travelling in the streamwise direction (Figure 12). In the present case, apart from the second roll being embedded in the general circulation (Figure 18a), no other convective features are visible in the transverse (xy) plane. Important differences also concern the intrinsic oscillatory nature of the flow. The recognizable convective vortices in the xz plane (Figure 18b) do not travel along the positive or negative z direction (as one would expect for a travelling wave); rather they assume the typical behavior of a standing wave, that is, the disturbances continuously oscillate between two limiting positions along the z-axis.

Given the considered value of the Prandtl number (Pr = 0.01) and the relatively low magnitude of the Rayleigh number for which transition to time-dependent flow occurs (Ra appreciably smaller than that required to excite hydrodynamic disturbances in the equivalent 2D case), these simple observations naturally lead to a fundamental question, that is, discerning the nature of the observed (3D) disturbances (Figure 18c).

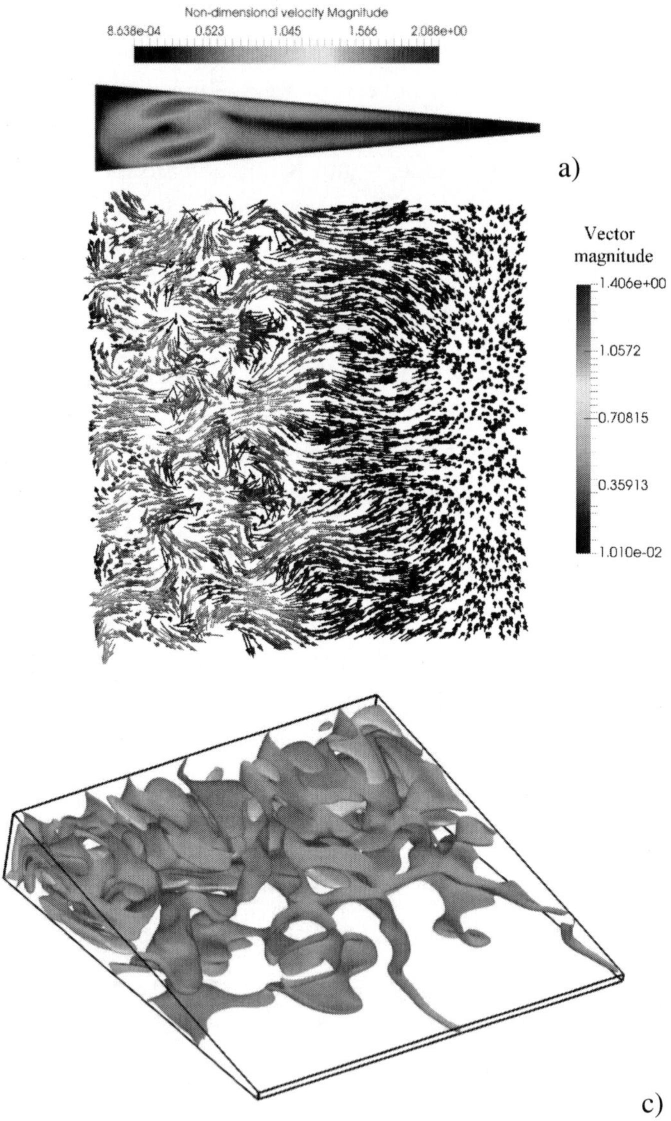

Figure 18. Snapshots of 3D oscillatory flow for Pr = 0.01, A_z = 10 (periodic boundary conditions, initial conditions corresponding to a quiescent thermally diffusive state), Ra = 2000 (Gr = $2x10^5$): a) velocity distribution in the xy plane, b) Velocity vector plot in the xz mid-plane, c) isosurfaces of the *w* (along z) component of velocity.

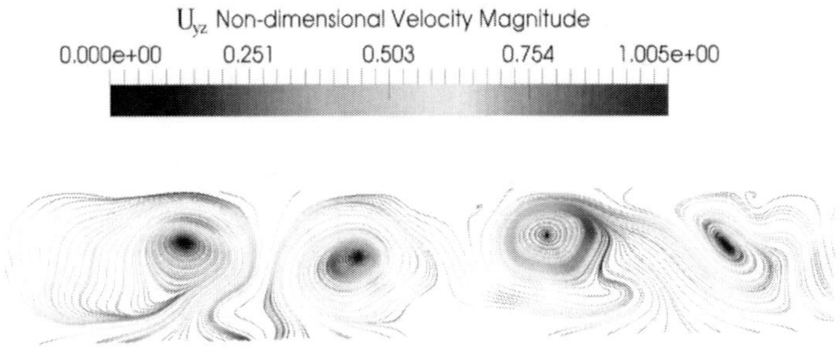

Figure 19. Projection of the velocity field in the yz plane at x = 1/8A_x (Pr = 0.01, A_x = 10, A_z = 10, periodic boundary conditions, Ra = 2000, temperature equation uncoupled).

Resolving this question, however, is not straightforward as one would imagine as it requires separating different contributions, which are generally intertwined.

Interestingly, Lappa and Ferialdi (2018b) showed that direct numerical solution of the overarching eqs. (1)-(3) (Sect. 1.3) can still be used as a valid tool to get meaningful information on these phenomena provided it is used in combination with specific rational arguments about the hydrodynamic and hydrothermal fundamental modes of instability and related differences (see again Sects. 1.1 and 1.2).

More precisely, they elaborated a specific analysis strategy based the possibility to uncouple the energy and momentum equations (solving the balance equations for mass and momentum only while assuming a fixed thermally diffusive distribution for the temperature).

The physical reasoning underpinning this approach can readily be understood by taking into account that the simple action of preventing the temperature field from developing changes should impede the onset and ensuing growth of any disturbance of hydrothermal nature.

The outcomes of such simulations for the same conditions considered for Figure 18 (shown in Figure 19) reveal that, as four longitudinal rolls (with axes parallel to the applied temperature gradient, i.e., oriented along the x axis) are still present (and since any processes that depend on the amplification of the aforementioned helicoidal disturbances are excluded in this case), the oscillatory dynamics reported in Figure 18 must therefore be considered of a hydrodynamic nature (which is relatively an unexpected result if one considers that hydrodynamic modes are generally thought of as purely two-dimensional disturbances).

Interestingly, Lappa and Ferialdi (2018b) could even track precisely the process leading from the initial steady state shown in Figure 17 to the much more involved situation evident in Figure 18 through evaluation of the so-called *correlation dimension*, which can generally be regarded *as a measure of the degrees of freedom possessed by a system.*

A rigorous definition of this quantity can be found in the literature where it is introduced as a generalization of the classical concept of dimension of a subset of the Euclidean space. While the dimension of a subset of the Euclidean space can be seen as the number of parameters needed to locate precisely a generic point of the set, the correlation dimension somehow accounts for the rate with which the typical 'orbit' (described by a system in the space of phases) 'visits' different parts of an attractor (Grassberger, 1983). Most remarkably, this percentage can take non-integer values. Following Balatoni and Renyi (1956), assuming the phase space to be partitioned via a grid of cubes *C* having characteristic size r (side length), the correlation dimension can formally be expressed as:

$$\mathcal{D}_2 = \lim_{r \to 0} - \frac{\ln\left\{\sum_j \left[\tau(C_j)\right]^2\right\}}{\ln(r^{-1})} \tag{6}$$

where C is the generic cube with side having length r in the phase space, $\zeta(\zeta_o,t)$ is the orbit originating from a generic initial condition ζ_o, and $\tau(C, \zeta_o, \Delta t)$ formally represents the percentage of time that $\zeta(\zeta_o,t)$ spends in C in a time interval Δt (Lappa, 2009). Over the years, algorithms able to provide an estimate of this quantity have been conceived, a relevant example being represented by the procedure defined by Grassberger and Procaccia (1983ab).

This dimension is generally regarded as an effective measure of the number of 'active modes' in a dynamical system (Theiler, 1986). Its values computed with the algorithm by Grassberger and Procaccia (1983ab) for the cavity shown in Figure 18 and magnitudes of the Rayleigh number progressively increasing from 1000 towards 2000 are reported in Figure 20, from which the reader will realize how quick the progression towards chaotic features can be in such a restricted interval of the control parameter when the flow is allowed to be 3D.

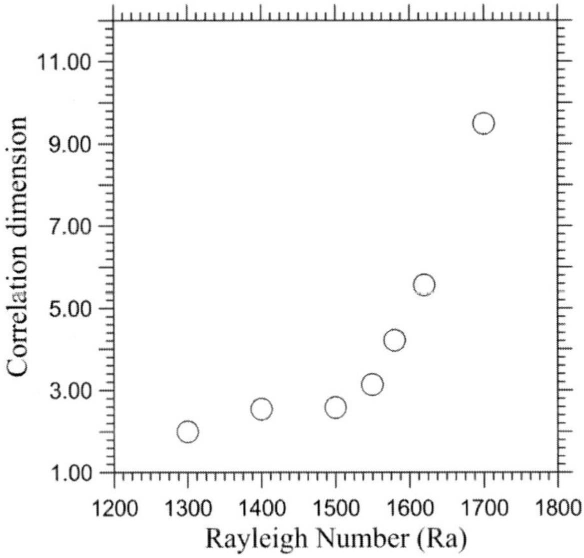

Figure 20. $\mathcal{D}_2$ versus Rayleigh number (Pr = 0.01, A_x = 10, A_z = 10, η = 0.1, periodic boundary conditions).

More precisely, as soon as Ra becomes larger than 1550, $\mathcal{D}_2$ exceeds 3 ($\mathcal{D}_2 \cong 3.14$ for Ra = 1550, $\mathcal{D}_2 \cong 4.22$ for Ra = 1580) and then suddenly increases for higher values of Ra. Such a gain in chaotic features can clearly be interpreted as the quick development of high-dimensional chaos. The reader may come to the same conclusion by taking a look at Figure 21 (showing the sequence of 'signals' for increasing values of Ra) or directly at the frequency spectra reported in Figure 22 for Ra = 2000. The latter figure is particularly meaningful as it provides immediate evidence for the rather *turbulent nature* of the 3D flow established for small values of the Rayleigh number when the dominant disturbances are hydrodynamic. This should be regarded as another important distinguishing mark with respect to the 2D solutions.

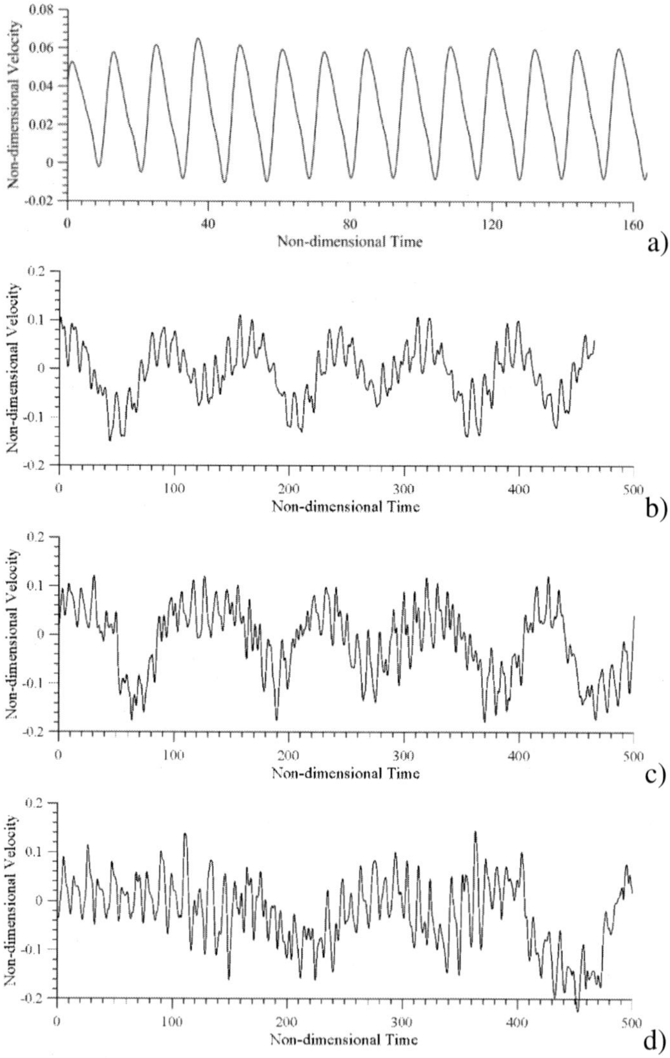

Figure 21. Velocity signals measured by a numerical probe located at a fixed point ($x = 3/8A_x$, $y = 0$, $z = A_z/2$) for $Pr = 0.01$, $A_x = 10$, $A_z = 10$, $\eta = 0.1$, periodic boundary conditions and different values of the Rayleigh number: a) $Ra = 1400$, b) $Ra = 1500$, c) $Ra = 1620$, d) $Ra = 1700$.

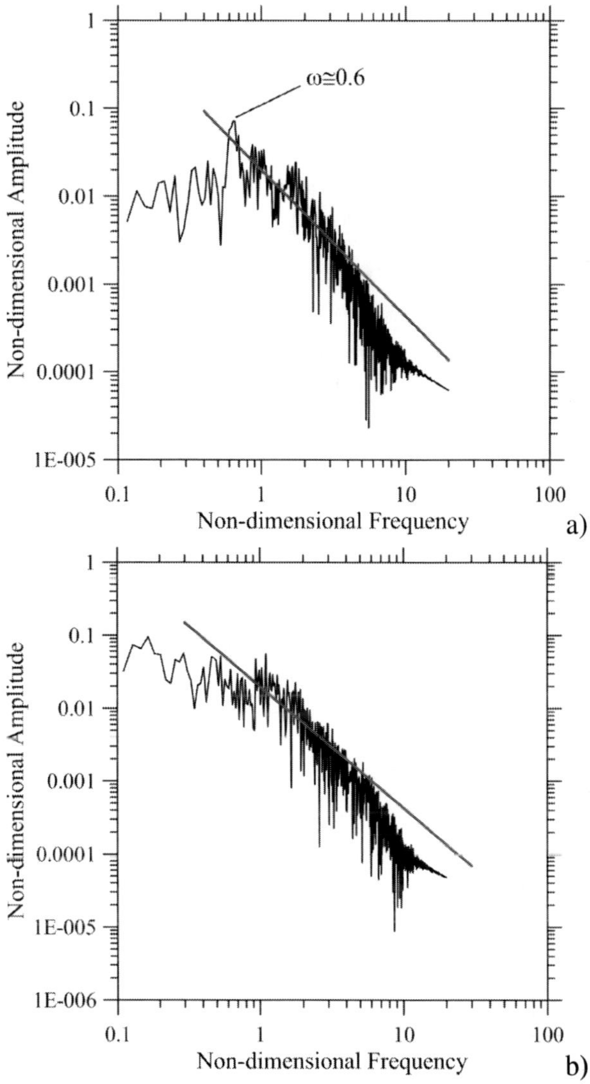

Figure 22. Non-dimensional velocity amplitude as a function of the non-dimensional frequency (Pr = 0.01, A_x = 10, A_z = 10, η = 0.1, periodic boundary conditions, Ra = 2000, Gr = $2\text{x}10^5$): a) temperature equation coupled (mixed regime with dominant hydrodynamic modes and OLR with $\omega \cong 0.6$), b) temperature equation uncoupled (purely hydrodynamic oscillatory mode of convection).

As discussed in Sect. 2.2.1, under the constraint of two-dimensionality (Figure 12) the flow retains an essentially laminar behavior. For instance, even for Ra = 9000 the velocity field cannot be considered turbulent as its frequency spectrum is limited to the presence of one or two incommensurate frequencies only. This behavior is in evident contrast with the scenario presented in Figure 22, where complex multimodal temporal dynamics can be seen even if the Rayleigh number is as small as 2000. The high-dimensional nature of the chaos relating to this state is confirmed by the fact that the spectra shown in such figure are in line with the typical behavior originally theorized by Kolmogorov (1941abc, 1942) about the development of isotropic turbulence. In order to make this specific aspect evident, following the same approach used by De et al. (2017), we have reported the frequency and related amplitude in Figures 22 using logarithmic scales for the axes and comparing the resulting diagram with a power law $P(\omega) = (\omega/\omega_c)^{-s}$ (where ω_c is a fitting parameter).

It can be seen that the frequency spectrum of velocities aligns perfectly with the $\omega^{-5/3}$ law in a wide interval of frequencies, as envisaged by the aforementioned Kolmogorov's turbulence similarity hypothesis. The related set of scales, also known as 'inertial range' corresponds to the so-called *energy cascade process* by which the directional biases of the large length scales are lost as energy is transferred to increasingly smaller scales. Such an interval is limited at one side by the typical length-scale L of the cavity, which is the scale at which energy is injected into the fluid via the action of buoyancy forces and at the other side by the scale where internal energy is produced by frictional effects at expenses of the kinetic energy of the flow.

Though, as illustrated in the preceding text, some interesting conclusions might be gathered from a careful comparison of 2D and 3D results, the variety of possible situations, however, can even be richer than what we have just discussed if the Prandtl number is allowed to change. According to numerical simulations (see again Lappa and Ferialdi, 2018b), indeed, an increase in the Prandtl number to Pr = 0.05 while keeping all the other parameters unchanged ($A_x = 10$, $A_z = 10$, periodic boundary conditions, Ra = 2000) can alter significantly the emerging dynamics. As evident in Figure 23, for such conditions, the pattern becomes much more regular in space. Moreover, the disturbances behave as *a wave travelling along the z direction* (Figure 24).

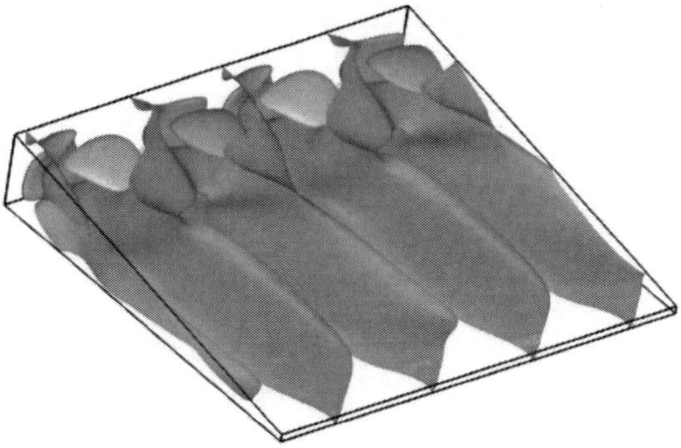

Figure 23. Isosurfaces (snapshot) of the *w* component of velocity ($Pr = 0.05$, $A_x = 10$, $A_z = 10$, $\eta = 0.1$, periodic boundary conditions, initial conditions corresponding to a quiescent thermally diffusive state, $Ra = 2000$, $Gr = 4x10^4$, OLR mode).

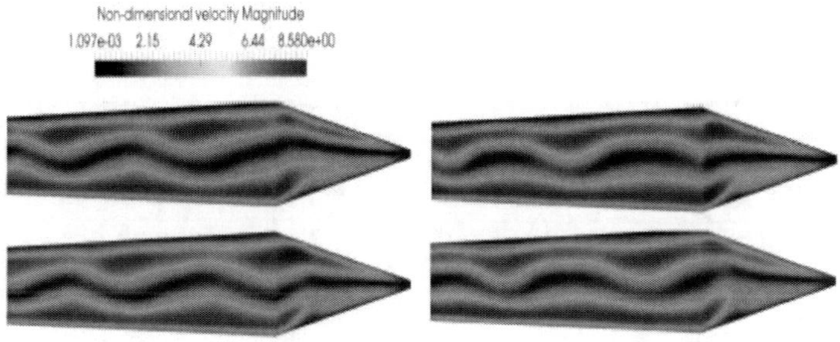

Figure 24. OLR mode ($Pr = 0.05$, $A_x = 10$, $A_z = 10$, $\eta = 0.1$, periodic boundary conditions, $Ra - 2000$, $Gr = 4x10^4$): four snapshots of non-dimensional velocity magnitude evenly distributed in time are shown (OLR with nondimensional angular frequency 2.81).

Simulations repeated after disabling the solution of the temperature equation (yet to exclude any disturbance of thermal buoyancy origin) have clearly indicated that the emerging solution can be regarded as an OLR mode as it *cannot survive after equation uncoupling*.

The outcomes of this study are summarized in Figures 25, which together with Figure 22, provide the spectrum of frequencies for the different situations examined so far, namely $Pr = 0.01$ ($A_x = 10$, $A_z = 10$ with periodic boundary conditions along z, $\eta = 0.1$) with the energy equation and the Navier-Stokes equations coupled and uncoupled (Figure 22a and 22b, respectively) and the analogous spectra for the case with $Pr = 0.05$ (Figures 25a and 25b).

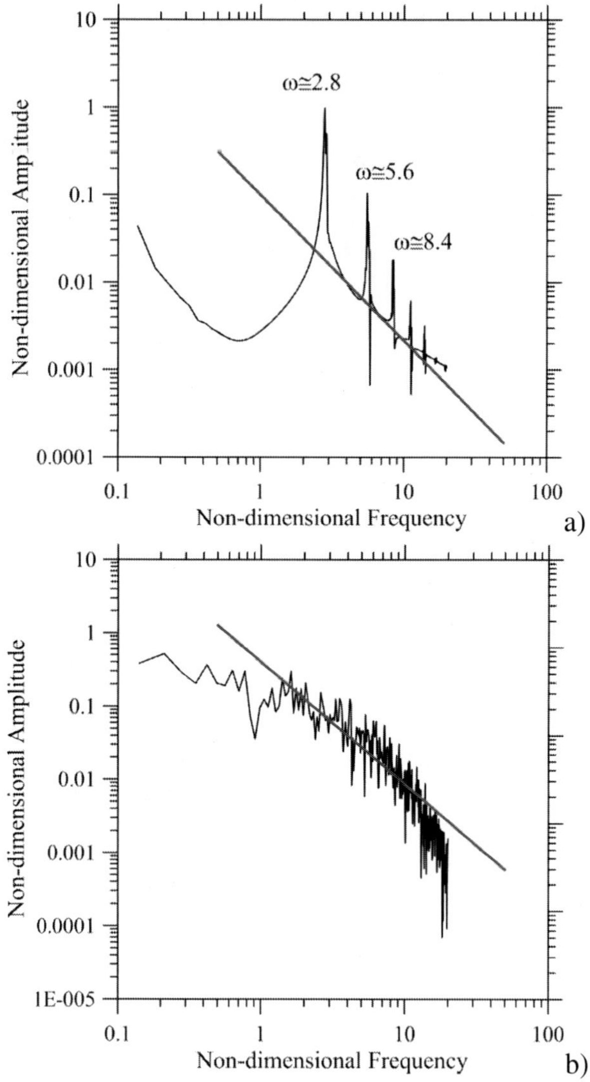

Figure 25. Non-dimensional velocity amplitude as a function of the non-dimensional frequency (Pr = 0.05, $A_x = 10$, $A_z = 10$, $\eta = 0.1$, periodic boundary conditions, initial conditions corresponding to a quiescent thermally diffusive state, Ra = 2000, Gr = $4x10^4$): a) temperature equation coupled (OLR mode), b) temperature equation uncoupled (hydrodynamic oscillatory mode of convection).

In particular, Figure 25 reveals that if, for Pr = 0.05, the onset of OLR modes is prevented by uncoupling the equations, the spectrum of frequencies takes the same complex structure obtained for Pr = 0.01 for the case of hydrodynamic disturbances.

Comparison between Figures 22a and 22b is also useful as it leads to the conclusion that though the spectrum retains a similar multi-frequency (broad-band) appearance when the temperature equation is switched off, the main peak visible in Figure 22a in the small-frequency region of the spectrum (around $\omega = 0.6$) disappears in Figure 22b, which should be regarded as a clear hint for the OLR nature of the perturbations being associated with it. Another interesting outcome of this figure is that it also definitely proves that hydrodynamic and OLR modes can *coexist*.

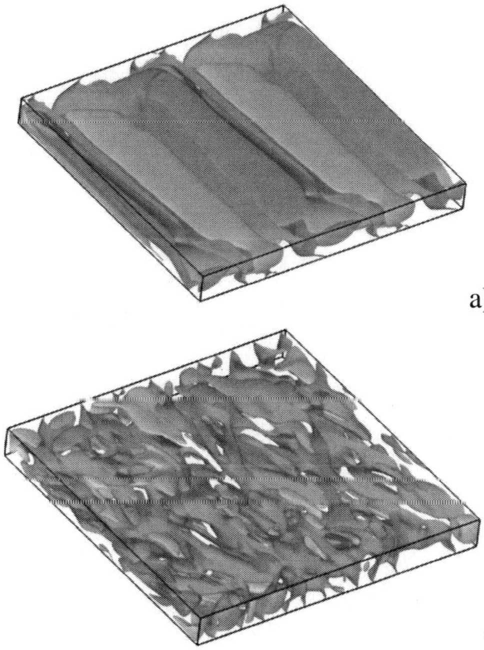

a)

b)

Figure 26. Isosurfaces (snapshot) of the w component of velocity (Pr = 0.05, $A_x = 10$, $A_z = 10$, $\eta = 1$, periodic boundary conditions, initial conditions corresponding to a quiescent thermally diffusive state, Ra = 2000, Gr = $4\text{x}10^4$): a) temperature equation coupled (OLR mode), b) temperature equation uncoupled (hydrodynamic oscillatory mode of convection).

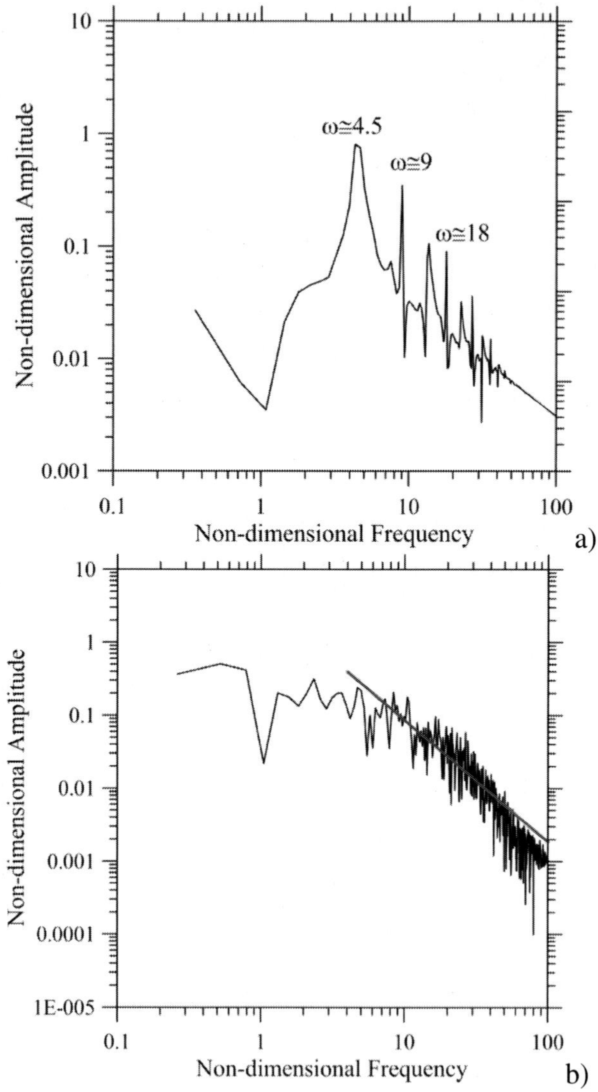

Figure 27. Non-dimensional velocity amplitude as a function of the non-dimensional frequency for the same conditions considered in Figure 26a and 26b, respectively.

The same conclusions could be drawn for the case with horizontal walls ($\eta = 1$, see Figures 26 and 27). The primary mode of convection is an OLR (Figures 26a and 27a) unless the energy equation is uncoupled (leading again to 3D hydrodynamic disturbances, Figures 26b and 27b).

Interestingly, cross examination of Figure 25a and 27a also indicates that the fundamental frequency of the OLR is slightly larger for the cavity with horizontal walls ($\omega \cong 2.8$ and $\omega \cong 4.5$ for $\eta = 0.1$ and $\eta = 1$, respectively).

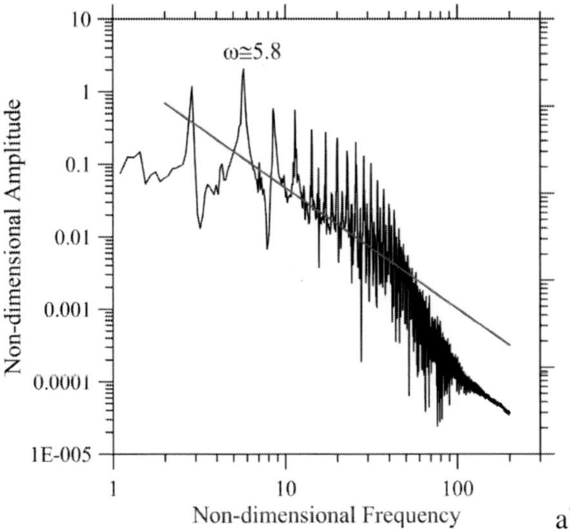

a)

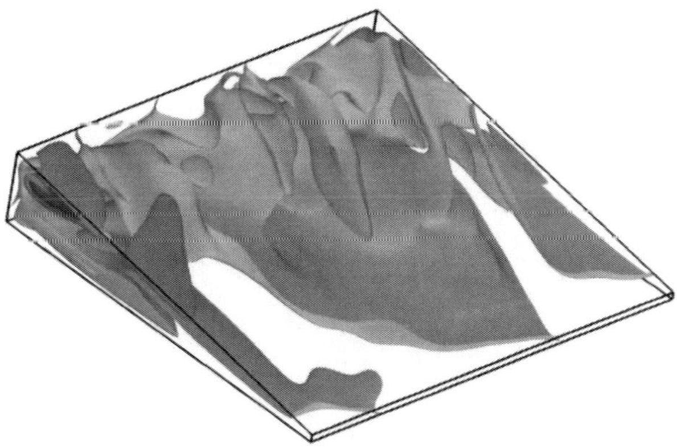

b)

Figure 28. Properties of the oscillatory state for Pr = 0.05, $A_x = 10$, $A_z = 10$, $\eta = 0.1$, periodic boundary conditions, initial conditions corresponding to a quiescent thermally diffusive state, Ra = 10000 (Gr = 2×10^5, mixed regime with hydrodynamic and OLR modes): a) frequency spectrum, b) isosurfaces (snapshot) of the *w* component of velocity.

These observations naturally lead us to re-introduce another of the interesting aspects being associated with these dynamics, that is, the problem relating to the existence of multiple solutions (already discussed to a certain extent for the 2D flow in Sect. 2.2.1). Multiple solutions are still possible when the constraint of two-dimensionality is removed. As an example, for Pr = 0.05 Lappa and Ferialdi (2018b) clearly observed different states existing independently at Ra = $2x10^4$.

Along these lines, yet for the case with converging walls and $\eta = 0.1$, Figure 28 shows the emerging flow (snapshot) and related frequency spectrum for Pr = 0.05 and Ra = 10^4 as a necessary pre-requisite to let the reader appreciate the multiple solutions that can be obtained when Ra is increased to $2x10^4$. For Ra = 10^4, a peak located at $\omega \cong 6$ in the left region (low-frequency interval) of the spectrum represents an OLR mode, while the right part (high-frequency interval) closely resembles that found for the case of dominant hydrodynamic disturbances. Therefore, this figure reinforces the idea that the two categories of disturbances can coexist (as also witnessed by the isosurfaces reported in Figure 28b, which have structure intermediate with respect to the disturbances visible in Figure 18c and Figure 23).

Figures 29 and 30, refer to the evolution of the field when the Rayleigh number is further increased. *Both relate to the same value of the Rayleigh number (Ra = $2x10^4$), the main difference being represented by the considered initial conditions*, namely, a quiescent thermally diffusive field for Figure 29 and the already developed velocity field (shown in Figure 28) for Figure 30. It can be seen that, while the flow in Figure 29 is essentially dominated by OLRs, this hydrothermal mode coexists with hydrodynamic disturbances in Figure 30. These findings are in line with the general idea that in a problem where coexisting branches of solutions are possible, the trajectories of the system in the space of phases can be attracted selectively to either of the solutions according to the considered initial state.

As a concluding remark, we wish to highlight that additional insights into the dynamics presented in the last part of this chapter may follow naturally from direct comparison of the frequency spectra with theoretical data available in the literature about the time scale limiting the inertial range of turbulence from below, i.e., the so-called Kolmogorov time scale (η_t).

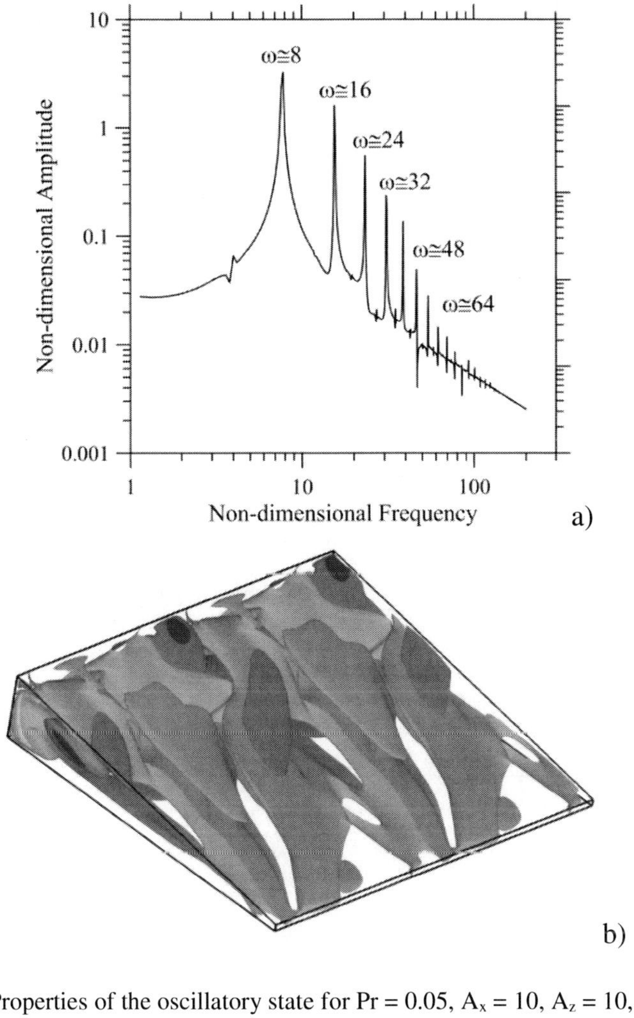

a)

b)

Figure 29. Properties of the oscillatory state for Pr = 0.05, $A_x = 10$, $A_z = 10$, $\eta = 0.1$, periodic boundary conditions, initial conditions corresponding to a quiescent thermally diffusive state, Ra = 20000 (Gr = $4\text{x}10^5$): a) frequency spectrum, b) isosurfaces (snapshot) of the *w* component of velocity.

Following Paolucci (1990) and Farhangnia et al. (1996), for a transversely heated cavity the Kolmogorov time and length scales can be evaluated, respectively, as:

$$\eta_t = 8\pi\left(\Pr Ra^3\right)^{-1/4} \tag{7a}$$

$$\eta_s = \pi \left(\frac{16\,\text{Pr}}{Ra} \right)^{3/8} \tag{7b}$$

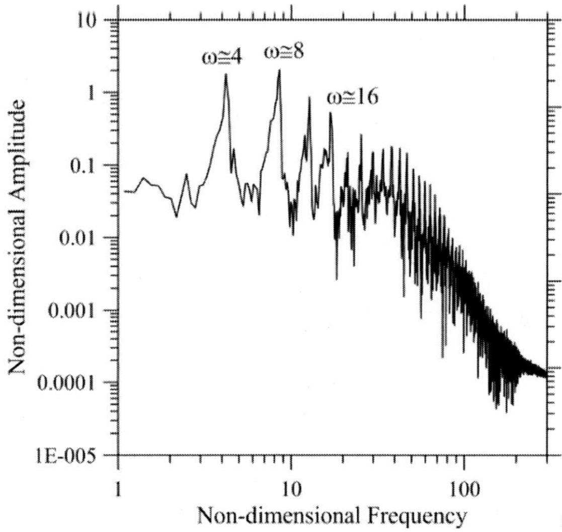

a)

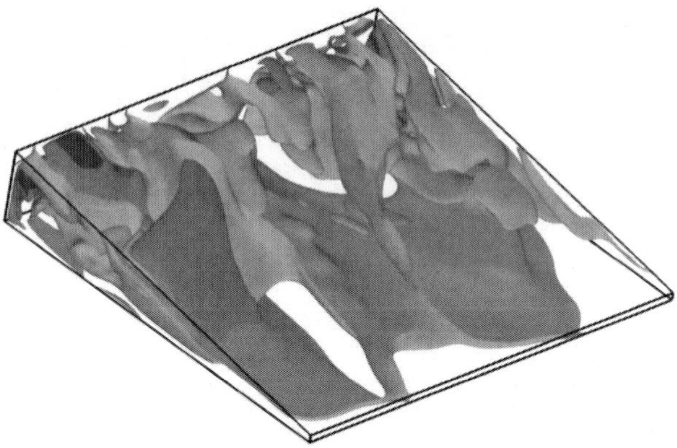

b)

Figure 30. Properties of the oscillatory state (mixed regime with hydrothermal and hydrodynamic modes) for Pr = 0.05, A_x = 10, A_z = 10, η = 0.1, periodic boundary conditions, Ra = 20000 (Gr = 4x10^5), initial conditions corresponding to Figure 28b: a) frequency spectrum, b) isosurfaces (snapshot) of the *w* component of velocity.

The good agreement between the right end of the intervals (shown in the preceding figures) where the Kolmogorov law is satisfied (for the case of 3D hydrodynamic disturbances) and the values provided by eq. (7a) led Lappa and Ferialdi (2018b) to the conclusion that these disturbances are the perturbations contributing more significantly to the transformative process leading these systems from low-dimensional chaos to fully developed turbulence.

CONCLUSION

The onset of typical instabilities in liquid metals undergoing buoyancy convection in differentially heated systems has been considered for different conditions and varying degrees of approximation. Starting from a review of results for infinite layers with horizontal or non-horizontal orientation produced in the past under the assumption of parallel flow and through the application of linear-stability-analysis techniques, we have considered configurations with progressively increasing complexity. These problems (relevant examples being represented by two-dimensional or three-dimensional cavities delimited by horizontal or non-horizontal solid walls) typically need to be addressed in the framework of advanced methods for the numerical solution of the overarching equations in their complete, time-dependent and non-linear form.

Initially, we have considered the typical behavior of these systems under the constraint of two-dimensional flow for which the emerging disturbances are expected to be purely hydrodynamic in nature (given the 3D nature of other perturbations such as the hydrothermal modes).

For the enclosures with converging or diverging walls and no net tilt, we have shown that three distinct regions of spatio-temporal behavior are possible in the considered range of compression/expansion ratios. The central one ($\eta \cong 1$) is the region of "standard" hydrodynamic modes emergence, that is the area of the space of parameters (η, Ra) where the Hadley flow is affected by the typical 'pulsating' disturbances (P) already observed in past studies for the case of perfectly rectangular cavities. The two external regions ($\eta < 1$ and $\eta > 1$), however, correspond to a different type of hydrodynamic disturbances, which spread continuously along the horizontal direction resembling 'travelling waves'. Though, for relatively small values of Ra, these two types of disturbances manifest separately, they are not truly progressive, nor are they reciprocally exclusive. These modes witness the existence of 'multiple' states of convection. In some cases (essentially in the high-Ra part of the space of parameters), disturbances relating to different branches can coexist leading to a flow in which a low-frequency (traveling) perturbation is modulated locally by a high-frequency (pulsating) disturbance.

When fully 3D configurations are considered, the third dimension enables the onset and growth of disturbances of hydrothermal origin (for which the possibility to propagate in the spanwise direction is an essential prerequisite). Even more interestingly, the presence of the third dimension has also a remarkable effect on the properties of the hydrodynamic modes, which can appear as 3D disturbances since the beginning.

Though hydrothermal waves seem to be more frequent as the Prandtl number is increased and they apparently rule out the development of hydrodynamic disturbances (for relatively small values of the Rayleigh number), however, hydrodynamic disturbances overwhelmingly re-enter the dynamics as soon as the momentum and energy balance equations are uncoupled or the Grashof number is sufficiently increased. Both modes can exist as multiple solutions and manifest separately or in combined form as revealed by a careful analysis of the frequency spectra.

We have also shown how, according to whether the emerging chaos displays a low-dimensional or high-dimensional nature, a relevant characterization of these systems can be based on the evaluation of the 'fractal' (correlation) dimension or the typical laws of the so-called Kolmogorov 'inertial range' of turbulence, respectively.

Future work shall be devoted to analyses fluid flow in 2D and 3D geometries with converging or diverging walls considering a net tilt with respect to the horizontal direction for which convective modes of Rayleigh nature, not possible for the conditions considered in the present chapter, can enter the dynamics and make the overall scenario even more complex.

REFERENCES

Afrid M. and Zebib A., (1990), "Oscillatory three-dimensional convection in rectangular cavities and enclosures", *Phys. Fluids*, 2(8): 1318-1327.

Balatoni J. and Renyi A., (1956), "On the notion of entropy", *Publ. Math. Inst. Hung. Acad. Sci.* 1, 9 (in Hungarian) [translation: Selected papers of A. Renyi, Vol. 1 (Budapest Academy, 1976) p. 558].

Braunsfurth M.G. and Mullin T., (1996), "An experimental study of oscillatory convection in liquid gallium", *J. Fluid Mech.*, 327: 199-219.

Chorin A.J., (1968), "Numerical solutions of the Navier-Stokes equations", *Math. Comput.*, 22: 745-762.

Crespo del Arco E., Pulicani P.P. and Randriamampianina A., (1989), "*Complex multiple solutions and hysteresis cycles near the onset of oscillatory convection in a Pr = 0 liquid submitted to a horizontal temperature gradient*", C. R. Acad. Sci. Paris 309, II: 1869-1876.

De A.K., Eswaran V., Mishra P.K., (2017), "Scalings of heat transport and energy spectra of turbulent Rayleigh-Bénard convection in a large-aspect-ratio box", *Int. J. Heat Fluid Flow*, 67: 111–124.

Delgado-Buscalioni R., (2001), "Convection patterns in end-heated inclined enclosures", *Phys. Rev. E* 64, 016303 17 pages.

Delgado-Buscalioni R., Crespo del Arco E. and Bontoux P., (2001), "Flow transitions of a low-Prandtl-number fluid in an inclined 3D cavity", *Eur. J. Mech. B/Fluids*, 329: 1-17.

Drazin P. and Howard L.N., (1966), "Hydrodynamic stability of parallel flow of inviscid fluid", *Adv. Appl. Mech.*, 9: 1-89.

Dupret F. and Van der Bogaert N., (1994), "Modelling Bridgman and Czochralski growth", in *Handbook of Crystal Growth* (ed. D.T.J. Hurle) 2: 877-1010. North-Holland, Amsterdam (1994).

Farhangnia M., Biringen S, Peltier L.J., (1996), "Numerical Simulation of Two-dimensional Buoyancy-driven Turbulence in a Tall Rectangular Cavity", *Int. J. Numer. Meth. Fluids*, 23(12): 1311 - 1326.

Gelfgat A.Yu., Bar-Yoseph P.Z. and Yarin A.L., (1999), "Stability of Multiple Steady States of Convection in Laterally Heated Cavities", *J. Fluid Mech.*, 388: 315-334.

Gershuni G.Z., Laure P., Myznikov V.M., Roux B., Zhukhovitsky E.M., (1992), "On the stability of plane-parallel advective flows in long horizontal layers", *Microgravity Q.*, 2(3): 141-151.

Gill A.E., (1974), "A theory of thermal oscillations in liquid metals", *J. Fluid Mech.* 64 (3): 577-588.

Grassberger P. and Procaccia I., (1983a), "Characterization of Strange Attractors", *Phys. Rev. Lett.*, 50: 346-349.

Grassberger P. and Procaccia I., (1983b), "Measuring the strangeness of strange attractors", *Physica D*, 9: 189-208.

Grassberger P., (1983), "Generalized dimensions of strange attractors", *Physics Letters A*, 97(6, 5): 227-230.

Gresho P.M., (1991), "Incompressible fluid dynamics: some fundamental formulation issues", *Ann. Rev Fluid Mech.*, 23: 413-453.

Gresho P.M., Sani R.T., (1987), "On pressure boundary conditions for the incompressible Navier-Stokes equations", *Int. J. Numer. Meth. Fluids*, 7: 1111-1145.

Hadley G., (1735), "Concerning the cause of the general trade winds", *Phil. Trans. Roy. Soc. Lond.*, 29, 58-62.

Harlow F.H. and Welch J.E. (1965), "Numerical calculation of time-dependent viscous incompressible flow with free surface", *Phys. Fluids*, 8: 2182-2189.

Hart J.E., (1972), "Stability of thin non-rotating Hadley circulations", *J. Atmos. Sci.*, 29: 687-697.

Hart J.E., (1983), "A note on the stability of low-Prandtl-number Hadley circulations", *J. Fluid Mech.*, 132: 271-281.

Hung M.C. and Andereck C.D., (1988), "Transitions in convection driven by a horizontal temperature gradient", *Physics Letters A* 132(5): 253-258.

Kengne J., Nguomkam Negou A., Tchiotsop D., Kamdoum Tamba V., Kom G.H., (2018), "On the Dynamics of Chaotic Systems with Multiple Attractors: A Case Study", In: Kyamakya K., Mathis W., Stoop R., Chedjou J., Li Z. (eds) Recent Advances in Nonlinear Dynamics and Synchronization. *Studies in Systems, Decision and Control, vol* 109. Springer, Cham.

Kolmogorov, A.N. (1941a) "The local structure of turbulence in incompressible viscous fluids at very large Reynolds numbers", *Dokl. Akad. Nauk. SSSR* 30: 299-303. Reprinted in *Proc. R. Soc. London* A 434: 9-13 (1991).

Kolmogorov A.N., (1941b), "On the degeneration of isotropic turbulence in an incompressible viscous fluids", *Dokl. Akad. Nauk. SSSR* 31: 538-541.

Kolmogorov A.N., (1941c), "Dissipation of energy in isotropic turbulence", *Dokl. Akad. Nauk. SSSR*, 32: 19-21.

Kolmogorov A.N., (1942), "Equations of turbulent motion in an incompressible fluid", *Izv. Akad. Nauk. SSSR ser. Fiz.* 6: 56-58.

Kuo H.P. and Korpela S.A., (1988), "Stability and finite amplitude natural convection in a shallow cavity with insulated top and bottom and heated from the side", *Phys. Fluids*, 31: 33–42.

Ladyzhenskaya O.A., (1969), *The Mathematical Theory of Viscous Incompressible Flow, Gordon and Breach*, 2nd Edition, New York - London, 1969.

Lappa M., (1997), "Strategies for parallelizing the three-dimensional Navier-Stokes equations on the Cray T3E", in *Science and Supercomputing at CINECA*, Vol. 11, pp. 326-340.

Lappa M., (2002), "Well-posed problems for the Navier-Stokes equations in the microgravity environment", *Microgravity & Space Station Utilization* (ISSN: 0958-5036), 3(4): 51-62.

Lappa M., (2009), *Thermal Convection: Patterns, Evolution and Stability*, John Wiley & Sons, Ltd (2009, Chichester, England).

Lappa M., (2019), "*Convective effects and traveling waves in transparent oxide materials processed with the floating zone technique*", this book.

Lappa M. and Ferialdi H., (2018a), Multiple solutions, Oscillons and Strange Attractors in ThermoViscoElastic Marangoni Convection, *Phys. Fluids*, 30(10), 104104 (19 pages).

Lappa M. and Ferialdi H., (2018b), "Gravitational Thermal Flows of Liquid Metals in 3D Variable Cross-section Containers: Transition from low-dimensional to high-dimensional chaos", *Chaos*, 28, 093114 (21 pages).

Lappa M. and Ferialdi H., (2017), "On the Oscillatory Hydrodynamic Instability of Gravitational Thermal Flows of Liquid Metals in Variable Cross-section Containers", *Phys. Fluids*, 29(6), 064106 (19 pages).

Laure P. and Roux B., (1989), "Linear and non linear study of the Hadley circulation in the case of infinite cavity", *J. Cryst. Growth* 97(1): 226-234.

Lin C.-C., (1944), "On the stability of two-dimensional parallel flows", *Proc. NAS*, 30(10): 316-324.

Monberg E., "Bridgman and related growth techniques". In *Handbook of Crystal Growth* (ed. D.T.J. Hurle), 2: 53-97. North-Holland, Amsterdam (1994).

Moukalled F., Mangani L. and Darwish M., (2016), *The Finite Volume Method in Computational Fluid Dynamics - An Advanced Introduction with Open FOAM and Matlab,* Springer International Publishing, 2016, New York).

Okada K. and Ozoe H., (1993b), "The Effect of Aspect ratio on the critical Grashof number for Oscillatory Natural Convection of Zero Prandtl Number Fluid: Numerical Approach", *J. Cryst. Growth*, 126: 330-334.

Okada K., and Ozoe H., (1993a), "Various computational conditions of oscillatory natural convection of zero Prandtl number fluid in an open boat heated and cooled from opposing vertical walls", *Numerical Heat Transfer*, Part A Applications, 23(2): 171-187.

Paolucci S., (1990), "Direct numerical simulation of two-dimensional turbulent natural convection in an enclosed cavity", *J. Fluid Mech.*, 215: 229-262.

Pratte J.M. and Hart J.E., (1990), "Endwall driven, low Prandtl number convection in a shallow rectangular cavity", *J. Cryst. Growth*, 102: 54-68.

Pulicani J.P., Del Arco E.C., Randriamampianina A., Bontoux P., and Peyret R., (1990), "Spectral simulations of oscillatory convection at low Prandtl number", *Int. J. Numer. Meth. Fluids*: 10(5), 481-517.

Rosenbluth M.N. and Simon A., (1964), "Necessary and sufficient conditions for the stability of plane parallel inviscid flow", *Phys. Fluids*, 7(4): 557-558.

Squire H.B., (1933), "On the stability of three-dimensional disturbances of viscous flow between parallel walls", *Proc. R. Soc. London, Ser. A*, 142: 621-628.

Temam R., (1968), "Une méthode d'approximation de la solution des équations de Navier-Stokes", *Bull. Soc. Math. France*, 98: 115-152.

Theiler J., (1986), "Spurious dimension from correlation algorithms applied to limited time-series data", *Physical Review A*, 34(3): 2427-2432.

Tollmien W., (1936), "General instability criterion of laminar velocity distributions", *Tech. Memor. Nat. Adv. Comm. Aero.*, Wash. No. 792 (1936).

Wakitani S., (2001), "Numerical study of three-dimensional oscillatory natural convection at low Prandtl number in rectangular enclosures", *J. Heat Transfer*, 123: 77-83.

In: Recent Studies in Materials Science
Editor: Patrick R. Lind
ISBN: 978-1-53615-270-8

Chapter 4

PHASE-STRUCTURAL TRANSFORMATIONS IN LOW-ALLOYED STEELS INTENDED FOR RAILWAY APPLICATIONS

V. G. Efremenko*[1,*]*, S. O. Kuz'min*[1]*, V. I. Zurnadzhy*[1]*,
***M. N. Brykov*[2]*, Yu. G. Chabak*[1] *and T. V. Pastukhova*[1]**

[1]Physics Department, Priazovskyi State Technical University,
Mariupol, Ukraine
[2]Welding Department, Zaporizhzhia National Technical University,
Zaporizhzhia, Ukraine

ABSTRACT

The present paper is devoted to the investigation of the thermo-kinetics features of overcooled austenite transformation in three grades of 0.75-0.77 wt.% C complex-alloyed (Mn-Si, Mn-Si-Cr-Mo) steel intended for rail applications. Magnetometric and thermal analyses were used to build Time-Temperature-Transformation and Continuous-Cooling-Transformation diagrams presenting the kinetics of austenite trans-

[*] Corresponding Author Email: vgefremenko@gmail.com.

formation in studied steels. Microstructural (LOM, SEM, TEM) techniques, XRD and microhardness measurements were employed to characterize the microstructure and the mechanisms of its formation. The cooling rate values for different microstructure type are determined for each studied steel. Different morphological types of bainite formed in steels depending on the transformation temperature (200 – 450°C) are observed and classified. The features of bainite fine substructure are described and discussed. The effects of the cooling rate and alloying elements on transformation characteristics are evaluated. The kinetics of austenite transformation in isothermal conditions is analyzed basing on Johnson–Mehl–Avrami equation and Cahn's theory.

Keywords:rail steel, phase transformation, TTT diagram, CCT diagram, austenite, pearlite, bainite, carbide

1. INTRODUCTION

Rails are very important structural elements which are critical for the reliability and safety of railway transportation. Higher requirements are formulated for rails to take into account the complexity of the exploitation conditions. The rail steels should provide higher complex of strength/hardness/wear resistance combined with increased ductility and impact toughness. Fine interlamellar pearlite (without proeutectiod ferrite/cementite) is usually considered as the optimal structure for rails, although low-temperature carbide-free bainite is also reported as a promising option for improving railway endurance. The microstructural status of steel is strongly influenced by its chemical composition and heat treatment. The study of solid-state phase-structural transformations in steel is needed to design the most suitable heat treatment approach for obtaining the desired rail microstructure and mechanical performance. The present work outlines the influence of the heat treatment profile on phase transformation in three grades of rail steels with different amounts of alloying elements. The main object of the work is to reveal the features of steel microstructure formation depending on the chemical composition and heat treatment schedule, including characterization of the microstructure obtained in different temperature ranges.

2. Mechanisms of Phase-Structural Transformations in Low-Alloyed Steels with near Eutectoid Carbon Content

Finding the optimal parameters of the mode of hardening heat treatment of steel rails involves the study of the characteristics of the transformation of overcooled austenite in a steel of a given chemical composition. It is known that during cooling from the austenitic temperature region a polymorphic $\gamma Fe \rightarrow \alpha Fe$ transformation occurs in the steel, and three transformation mechanisms are known as: reconstructive (pearlitic), displacive (martensitic) and bainitic, where the latter combines the features of the previous two. These mechanisms differ in the factor controlling the transformation kinetics: in pearlitic transformation, this factor is atoms diffusion, in the case of martensitic transformation this is shear resistance, and in the case of the bainite transformation the reaction is controlled by both factors. The kinetics of the polymorphic transformation of austenite is characterized by the presence of a maximum velocity at a temperature determined by the superposition of the effects of thermodynamic and diffusion factors [1].

Under isothermal conditions, pearlitic transformation usually takes place in the temperature range from point Ar_1 to the temperature corresponding to the "nose" of the C-shaped TTT diagram. However, there are experimental data revealing the pearlitic transformation below the "nose" of the diagram. Thus, the authors of [2] established the possibility of the existence of low-temperature pearlite, which forms at 375-325°C. The pearlite colony forms through the parallel growth of ferrite and cementite plates. Starting from the grain boundaries, the pearlitic transformation proceeds towards the centre of the grain due to the diffusion of carbon to austenite. The thickness of the forming plates depends on the mobility of carbon atoms [3]. The up-to-date conception of the mechanism of pearlitic transformation is based on fundamental works by Hull, Avrami, Johnson, Mehl, Zener, Cahn, Mirkin, Kolmogorov, etc.

Johnson–Mehl–Avrami equation [4,5] is often used to describe the kinetics of the pearlite transformation:

$$f = 1 - exp(-K\tau^n), \tag{1.1}$$

where f is the volume fraction of transformed austenite; τ is the isothermal holding time at a certain temperature, and K and n are kinetic parameters.

The kinetics of the pearlite transformation could also be described by the Mirkin–Kolmogorov equation [5, 6]:

$$V(\tau) = V_0\left[1 - exp(-\frac{1}{3}\pi nc^3\tau^4)\right], \tag{1.2}$$

where V_0 is the austenite initial volume fraction,%; n is the nucleation rate; c is the rate of nucleus linear growth.

The rate of growth of cementite plates in pearlitic colonies (G) is calculated according to the Zener–Hillert model by the expression [5]:

$$G = \frac{D_c^\gamma\left(C_{\gamma/\alpha} - C_{\gamma/Fe_3C}\right)}{2\Delta(C_{Fe_3C} - C_{\gamma/Fe_3C})}, \tag{1.3}$$

where D_c^γ is a coefficient of carbon diffusion in austenite; Δ - the interlamellar distance in pearlite; $C_{\gamma/\alpha}$ and C_{γ/Fe_3C} are the carbon contents in austenite at the austenite/ferrite and austenite/cementite boundary accordingly; C_{Fe_3C} is the carbon content in cementite.

The kinetics of the pearlitic transformation in alloyed steels (in particular, with the eutectoid carbon content) has been extensively studied in numerous works [7-9]. The simulation of the pearlitic transformation depending on various factors is considered in [10].

In some cases, the formation of pearlite proceeds by mechanisms which are different from the parallel growth of ferrite and cementite plates. In [71] it is shown that if austenite contains fine cementite particles or nucleation sites located at a distance of less than few micrometers from each other, the transformation is completed by the formation of spheroidal cementite inclusions in the ferrite matrix. This type of transformation is called "divorced eutectoid transformation". The formation of the granular-shaped cementite in eutectoid was also observed by the authors of [12, 13], who studied the pearlitic transformation of austenite subjected to preliminary cold deformation.

In steels alloyed with strong carbide-forming elements, pearlite transformation is a multistage process, starting with the precipitation of alloyed carbides from overcooled austenite [14]. Since proeutectoid ferrite is also oversaturated with carbon, carbides precipitate from the ferrite as well. Then, austenite undergoes transformation via eutectoid reaction, resulting in a mixture of carbides and oversaturated ferrite. At the final stage of the transformation, carbides precipitate already from eutectoid ferrite. It should be noted that carbides precipitated at different stages differ in shape: while the carbides precipitated from austenite are of compact shape, the carbides precipitated from ferrite are of fibrous elongated shape. It was found in [11] that during pearlitic transformation ferrite work-hardening is possible, which causes further recrystallization upon subsequent heating.

The application of external stresses exceeding the yield strength of austenite leads to the acceleration of austenite transformation; preliminary hot deformation has the same effect. The kinetics of austenite transformations is also significantly affected by high pressure: it was found that the kinetic curve of the pearlitic transformation shifts under the influence of pressure towards longer holdings, while the temperature of the maximum transformation rate decreases. Pearlite, formed at elevated pressure, has a coarser structure, with a larger interlamellar spacing. The slowdown of the pearlitic transformation under pressure is explained by lowering of the equilibrium temperature of the "austenite $\rightarrow$ ferrite+carbide" reaction, i.e., by decrease in the thermodynamic stimulus with a decrease in atoms diffusion mobility.

Bainitic transformation occurs at temperatures where diffusion mobility of iron and alloying elements are very low while carbon atoms mobility remains rather high. The generally accepted vision of the mechanism of bainitic transformation is described as follows: "...transformation, at least in the initial stages, can be imagined to consist of a preparatory diffusion process of redistributing carbon inside austenite and subsequent lattice rearrangement in those carbon-depleted areas for which the bath temperature is a Ms point" [15]. Thus, before the start of bainitic transformation, the austenite becomes heterogeneous in carbon content.

It has been established that bainitic transformation is directly related to the redistribution of carbon in austenite. Comparison of the activation energy of α-phase crystals growth (50.4-71.4 kJ/mol) with the activation energy of carbon diffusion (134.4-151.2 kJ/mol) gave a reason to assume that carbon redistribution in austenite during bainitic transformation cannot proceed through normal diffusion [16]. According to Lyubov and Aleksandrov [16], the removal of carbon atoms from the front of a growing ferritic crystal occurs due to the effect of elastic stress fields at the phase boundary, leading to relaxation of these stresses.

According to the theory of spinodal decomposition [17], with a decrease in the transformation temperature, on the curve of free Gibbs energy there appears a segment with a negative curvature ($\frac{\partial^2 G}{\partial X_C^2} < 0$). As a result, at temperatures below the pearlitic range, austenite spontaneously begins to stratify into areas with different carbon concentration. In carbon depleted areas, austenite is transformed into α-phase by a displacive mechanism with further carbide precipitating from the α-phase. In carbon enriched areas, carbides first precipitate from austenite, leading to increase of the Ms point with further austenite transformation into α-phase through a displacive mechanism; the latter as well undergoes decomposition with carbides precipitation. Carbon-enriched austenite in certain areas is very stable to shear transformation retaining in the structure until the end of the transformation; it can partially transform into martensite when cooled to room temperature. Thus, bainite is a complex structure conglomerate consisting of ferrite of different oversaturation with carbon, a carbide phase and carbon-rich retained austenite.

A characteristic feature of bainitic transformation is that the formation of special carbides during the transformation is not observed even for steels alloyed with strong carbide-forming elements. This is only cementite carbide, which precipitates during bainitic transformation, while the content of carbide-forming elements in cementite does not differ from the average content in steel. A slight enrichment of cementite with these elements can be observed only in the upper region of the bainitic transformation, where the superposition of pearlitic and bainitic transformations takes place. This fact is explained [18] by the absence of diffusion of alloying elements in austenite during the bainitic transformation.

The author of [18] revealed that the difference in the mechanisms of formation of upper and lower bainite in eutectoid steel also lies in the phase nucleation mechanisms: in the first case, the transformation simultaneously exhibits the signs of both diffusion and shear mechanisms, while in the second case it is fully shear. In addition, there are differences in the mechanism of the interfacial boundary motion during the transformation.

The data presented in [19-21] show that there is a variety of morphological types of bainite, determined by the composition of particular steel and the temperature of bainitic transformation. Currently, there is no generally accepted classification of the types of structures formed in steels at temperatures below the pearlite transformation range. There are several classifications adopted for bainite structures. According to one of them [18], there are: a) nodular bainite, b) columnar bainite, c) upper bainite, d) lower bainite arising along grain boundaries (grain boundary allotriomorphic) and e) inverse bainite. Experts of "*ISIJ* Bainite Committee" proposed another list of ferrite types presented in bainite such as: a) polygonal bainite, b) quasi-polygonal bainite, c) Widmanstätten bainite, d) granular bainite, e) bainitic ferrite and f) cubic martensite [22].

In [23], it was found that in low-carbon low alloyed steels a granular bainite is formed under continuous cooling. This kind of bainite is a mechanical mixture of α-phase, globular carbides and globular areas of retained austenite; the α-phase consists of plate or lath ferrite and twinned martensite.

Articles [19-21] provide a classification of the types of bainitic structures in steels of different carbon content with the corresponding mechanisms description. It is proposed that upper bainite is subdivided into low-carbon upper bainite and high-carbon upper bainite. According to [19-21], low-carbon upper bainite is formed at 550-450°C in steels containing not more than 0.7 wt.% C. This structural component is the plates of supersaturated ferrite or products of its tempering, between which thin films of martensite or carbon-rich austenite are revealed. The mechanism of formation of this structure is supposedly ferrite nucleation along the grain boundaries as parallel plates. The ferrite plates increase in size, rejecting carbon into the austenite. Upon cooling, enriched austenite transforms into martensite or pseudo-eutectoid, or it remains in the structure as retained austenite.

High-carbon upper bainite usually has a specific feathery morphology. Its laths are a conglomerate of carbide and α-phase separated by the layers of martensite/retained austenite. The formation of high-carbon upper bainite occurs due to alternating carbide precipitation and transformation of carbon-depleted austenite into α-phase. Aside from feathery upper bainite, the author of [90] proposes the concept of dendritic and comb-shaped high-carbon upper bainite, observed in steels alloyed with carbide-forming elements.

According to [20], the lower bainite should also be subdivided into low-carbon and high-carbon lower bainite. The first one appears at temperatures below the Ms point, being a product of martensite tempering. The second one is formed at temperatures lying above the Ms point as a result of the redistribution of carbon in austenite. High-carbon lower bainite has a feathery structure which is a conglomerate of carbides and supersaturated ferrite alternating with layers of retained austenite. This morphology indicates the pulsating, discrete nature of the bainite transformation, which consists of numerous cycles of carbon redistribution and ferrite formation [20].

Alloying elements have a significant effect on the kinetics of austenite transformations. By the character of this influence, the alloying elements are usually divided into two groups. The first group includes Ni, Si, Cu, Co, and Al, which only quantitatively effect on C-shaped curve on the TTT diagram. All of these elements, except cobalt, increase the incubation period, as well increasing the total duration of austenite transformation [24, 25]. The second group includes carbide-forming elements (Mn, Cr, Mo, W, V, etc.), which increase the stability of austenite for pearlitic transformation while splitting C-shaped curve into pearlitic and bainitic domains.

As a rule, alloying elements inhibit eutectoid transformation of overcooled austenite. This is due to the factors outlined in [25] as: (a) alloying elements reduce the thermodynamic stimulus of the process of rearrangement of iron atoms at the phase interface during the $\gamma Fe \rightarrow \alpha Fe$ transformation; (b) alloying elements reduce the diffusion mobility of carbon in austenite, thus obstructing the redistribution of carbon atoms ahead of the growing pearlitic colony; (c) in the presence of carbide-forming elements, the pearlitic transformation should proceed through the formation of special (alloyed) carbides, which requires diffusion redistribution of carbide-forming elements atoms. Taking into account the lower diffusion coefficients of these elements, the formation of special carbides at the temperature of bainitic domain is kinetically unfavourable as compared with unalloyed cementite formation.

Manganese, chromium, and silicon increase the stability of overcooled austenite in both the pearlitic and bainitic domains. Molybdenum and tungsten, which have a slight effect on the rate of $\gamma Fe \rightarrow \alpha Fe$ polymorphic transformation, practically do not effect bainite transformation [24], while significantly inhibit the pearlitic transformation due to the formation of special carbides. In the case of alloying with different alloying elements, a synergy is observed which exceeds the additive effect of their individual influence on austenite stability [25].

Manganese and silicon are the most widely used alloying elements. They are used as steel deoxidizers, as well as for improving the mechanical properties of steel. It is known that Mn and Si are involved in the solid-solution strengthening of ferrite, wherein silicon by three times is more effective ferrite strengthener than manganese [26]. Manganese strongly increases the austenite stability in pearlitic temperatures, and this manganese effect is twice as high as chromium and by 1.5 times than molybdenum [25]. Also, manganese is the most effective element (excluding carbon and nitrogen) for reducing the Ms temperature, resulting in an increase in the amount of retained austenite in steel [25].

Manganese is one of the main alloying elements used for rail steels having near-eutectoid carbon content. Examples are the rail steels K76, K76T, K76V, etc. according to Russian standard 51685-2000 with a content of 0.71-0.82 wt.% C and a manganese content of 0.75-1.15 wt.%. The TTT diagram for these steels has one kinetic maximum observed in the temperature range of pearlitic transformation. Troostite/sorbite, with fine interlamellar spacing formed at the temperature close to the "nose" of the TTT diagram, is considered to be the optimal microstructure for rails [27]. Despite the fact that increasing carbon content inhibits the formation of proeutectoid ferrite, it is not possible to completely avoid proeutectoid epitaxial ferrite in the rail under transformation at higher temperatures 620-630°C. The formation of epitaxial ferrite could be prevented by lowering the temperature of austenite transformation (up to 600°C) by using accelerated cooling. During spraying the rail head with air–water mixture, austenite is transformed at 450-500°C leading to troostite/sorbite microstructure without proeutectoid ferrite [28, 29].

With an increase in the manganese content from 0.30 to 1.32 in steel with 0.60-0.65 wt.% C, the time of austenite transformation in the pearlitic temperature range increases from 1 to 60 sec [29], while the pearlitic domain is shifted to a lower temperature range. However, the increase in manganese content may cause the formation of a heterogeneous microstructure with the bainite or martensite areas, which are unacceptable for the rail head.

The available literature data on the kinetics of overcooled austenite transformation in rail steels differ significantly. Thus, according to [29], the minimum stability of austenite is referred to the range of 600-500°C, while in [30, 31] the "nose" of C-shaped diagram was revealed at 520-480°C. In [32] it was found that in rail steel containing 0.76 wt.% C and 0.98 wt.% Mn pearlite structure of different dispersion is formed under cooling at a rate less than 32 K/sec; when cooled at a rate of 32 K/sec, the resulting structure consists of a mixture of thin lamellar pearlite (with an interlamellar distance of 0.4 μm) and bainite areas. A fully martensitic structure is formed when cooled at a rate of over 60 K/sec. In contrast, the authors of [33] present data confirming that the critical quenching rate of rail steel for martensite is in the range of 14-20 K/sec. Such a big scatter of values implies a need for additional studies in order to clarify the kinetics of transformation in rail steel.

In recent decades, new generation of rail steels have been actively developed. These are eutectoid Mn-Cr-Si-Mo-V(Nb) steels and steels with a carbide-free bainite structure. The former [33] provide the same microstructure in the rail head as mentioned above for Mn-Si steels (K76, K76T, K76V, etc.), that is, fine lamellar troostite/sorbite, while they ensure a larger depth of hardening over the head cross-section. The second ones are outcome of Bhadeshia with co-workers research dedicated to creating of nano-structured steel with carbide-free bainite (Super Bainite) [34]. Bainitic steels for rails provide a bainitic structure over the entire rail cross-section, which, according to [35-37], ensures a higher complex of mechanical properties compared with troostite/sorbite structure. Bainitic rails require reduced maintenance and increased safety combined with higher wear resistance as compared to a pearlitic rail structure. Bainitic rail steels contain a reduced amount of carbon (0.25-0.4 wt.%) with a higher content of alloying elements (1.5-3.0 wt.% Mn, 1.5-2.0 wt.% Si, 1.0- 2.0 wt.% Cr, up to 0.5 wt.% Mo, up to 0.2 wt.% V). The kinetics of phase transformations in bainitic steels has been studied quite well, with the construction of both TTT and CCT diagrams [37]. Similar data concerning Mn-Cr-Si-Mo rail steels are rarely found in the literature [33, 38], that requires additional studies.

3. METHODS

Hot rolled bars made of steel 75Mn, 75Mn2Si and 75MnCrSiMo were received from metallurgical enterprise as research materials. The chemical composition of steels is presented in Table 3.1. The samples were obtained by cutting from hot rolled bars. Austenitization of the samples before cooling and isothermal treatment was carried out in laboratory furnaces at 850°C. Isothermal holding was performed in a salt bath in a melt of K_2NO_3 and $NaNO_2$ mixture.

The kinetics of austenite transformation was investigated using a magnetometer equipped with a strain gauge system for fixing the magnetization of the sample. The sample with dimensions of 2×10×12 mm after austenitization for 5 min was transferred to the salt bath positioned between the poles of the magnet. As the $\gamma Fe \rightarrow \alpha Fe$ transformation developed, the sample was drawn into the magnetic field of the magnetometer, deforming the beam with the strain gauges. The signal from the strain gauges was recorded on a computer using an analog-to-digital converter.

By processing the magnetometric signal, the kinetic curves of austenite transformation were constructed. The magnetometric signal obtained from the annealed sample was adopted as the standard signal corresponding to 100% of transformation. The beginning of the transformation was considered the appearance of 2–3 vol.% of the ferromagnetic phase (αFe). The results of the magnetometric analysis were checked and supplemented by microstructural analysis of the samples cooled in water after being removed from the bath.

Investigation of the processes of phase transformation under continuous cooling was carried out using a thermal method. The samples of round section of 20 mm length were austenitized at 850°C with subsequent cooling at different cooling rates. The cooling rate was varied by changing the diameter of the samples (1.5-5 mm), due to insulation with quartz or asbestos, as well as through blowing with fans of different capacities; this allowed the cooling rate to be varied from 0.03 to 80 K/sec (these values are averaged for the temperature range of 800- 200°C). The temperature was recorded using a personal computer equipped with an "Advantech" analog-to-digital transformation card. The start temperature of the phase transformation was determined by the position of the bending points on the cooling curve. After cooling, the samples were subjected to microscopic analysis and to microhardness measurement in order to reveal their structural status. When studying transformation, 3–4 samples were used for each experimental point, with subsequent averaging of the results.

The microstructure was studied using Axiovert 40 MAT (Carl Zeiss) optical microscope and Ultra 55 (Carl Zeiss) scanning microscope. The samples were prepared in accordance with the standard preparation procedure and further etched with a 4% nital solution. Transmission electron microscopy (TEM) study was conducted using a JEM-100-C-XII (JEOL) microscope with the accelerating voltage of 100 kV. Thin foils were used, which were cut from the samples subjected to different heat treatment regimes. The method of foil preparation consisted in mechanical thinning to a thickness of 0.10-0.15 mm, followed by cutting (using an electric spark installation) of disks with a diameter of ~ 3 mm. The disks were subjected to electrolytic jet polishing using the "TENUPOL-2" unit in the electrolyte composed of 40 ml of perchloric acid, 400 ml of methyl alcohol, and 240 ml of butyl alcohol. The microhardness was determined using the "Affri DM8" unit at a load of 50 g. The results presented are averaged from 7–10 measurements made on the same microstructure area.

4. Structure Formation under Isothermal Subcritical Treatment

The transformation of overcooled austenite in steel 75Mn under isothermal treatment was studied in subcritical temperature range. The austenized samples were held at constant temperature varying in the range of 200-650°C with increments of 50 K. As a result of the studies, the kinetic curves of transformation were obtained (Figure 4.1a). Using these curves, the TTT diagram for steel 75Mn was constructed, shown in Figure 4.1b. As can be seen, the classical C-shaped TTT diagram was obtained for steel 75Mn, showing one kinetic maximum of the transformation, corresponding to a fairly wide temperature range of 400-550°C, in which the values of the incubation period (τ_{inc}) varies negligibly. The minimum stability of austenite corresponds to t_{is}=450-500°C (τ_{inc}=4 c). The transformation in 75Mn steel was observed even at 225-250°C, which lies below the martensitic point Ms (260°C).

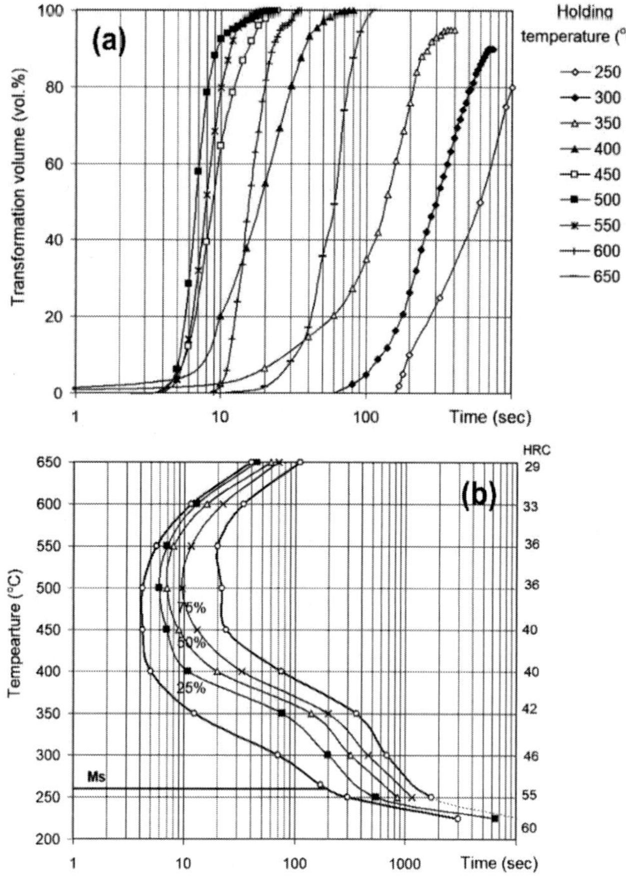

Figure 4.1. (a) Transformation kinetic curves and (b) TTT diagram for steel 75Mn.

As follows from Figure 4.2, under holding at 225-250°C, the transformation begins with the formation of dark lens-like laths (located arbitrarily, irrespective of the grain boundaries), which are clearly distinguished on the light martensite/austenite background. The laths are oriented along certain crystallographic planes and resemble coarse needle martensite. The development of the transformation took place mainly due to the appearance of new "needles" (laths), since the "needles" that had arisen earlier did not significantly grow in size. The austenite transformation finishes at 225-250°C with the formation of a dense "forest" of interweaving acicular laths; some of the laths, when etched, acquire a lighter contrast, standing out against the background of darker "needles" (Figure 4.2, c). This microhardness of this microstructure is 600-680 HV.

Table 3.1. Chemical composition of studied steels

Steel grade	Content (wt.%)						
	C	Si	Mn	Cr	Mo	S	P
75Mn	0.75	0.28	0.94	0.02	≤ 0.01	0.018	0.020
75Mn2Si	0.76	0.42	1.35	0.02	≤ 0.01	0.011	0.018
75MnCrSiMo	0.77	0.46	0.98	0.48	0.08	0.022	0.025

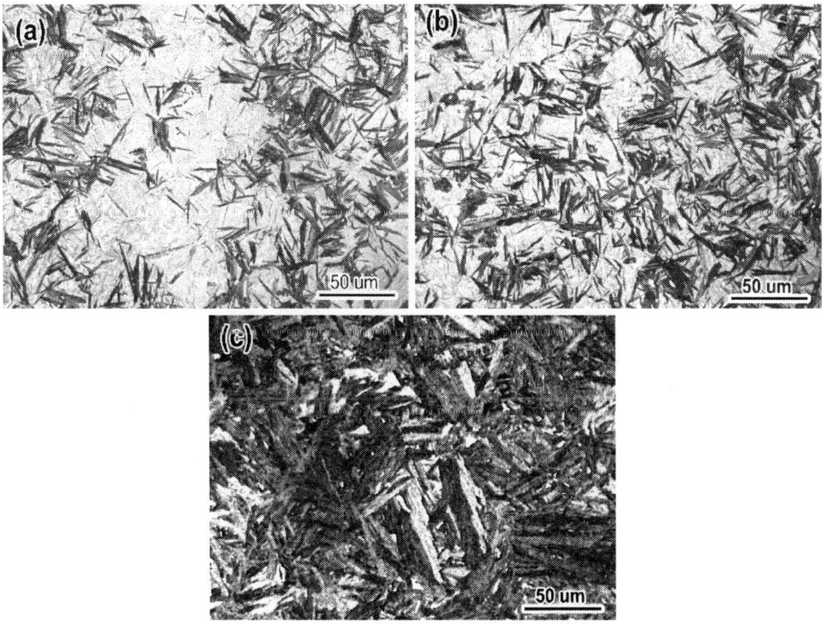

Figure 4.2. Microstructure of steel 75Mn obtained by holding at 250°C for: (a) 10 min, (b) 15 min, and (c) 30 min.

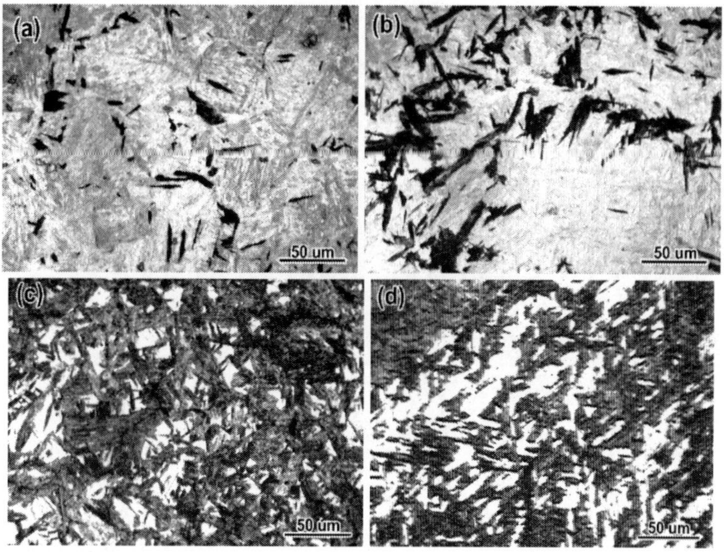

Figure 4.3. Microstructure of steel 75Mn obtained by holding at 300°C for: (a) 1,5 min, (b) 4 min, (c) 15 min and (d) 30 min.

When the holding temperature is 300°C (i.e., higher than Ms point), the transformation products undergo certain changes relative to 250°C. As follows from Figure 4.3, the transformation begins mainly from the boundaries of austenitic grains. The transformation products are shaped as elongated dark plates oriented along crystallographic planes, but to a lesser extent resemble needle-like martensite. The transformation develops through the formation of new plates and the growth (in the side and frontal directions) of the previously formed plates. With increase in the holding duration, the plates thicken, merging into packets. Gradually, the boundaries of the packets and individual plates become blurred, and the expanse between them is filled with heterogeneous light-grey areas having a certain crystallographic orientation. This leads to the formation of "chequered" microstructure with alternating light and dark micro-areas (Figure 4.3, d). The same character of structure formation is noted for holding at 350°C (Figure 4.4). The micro-hardness of bainite formed at 300-350°C is significantly reduced relative to 225-250°C (412-473 HV and 600-680 HV respectively).

With an increase in the holding temperature to 400°C, a radical change in the transformation mechanism and the structure morphology occurs. After short exposures (5-6 sec), dark rounded colonies of eutectoid (troostite) (Figure 4.5) are detected along the austenitic grains boundaries. These colonies further grow towards the centre of the grain. In some places, a thin network of epitaxial ferrite is formed along the grain boundaries. At t_{is}=400-450°C, the microhardness of the steel is approximately the same as that estimated as 374-389 HV. At higher holding temperatures, the eutectoid structure (Figure 4.6) is formed in steel 75Mn with microhardness decreasing from 356 to 292 HV proportionally to the increase of the holding temperature. At 600-650°C the structure consists of dispersed eutectoid (sorbite) with occasional areas of coarse pearlite.

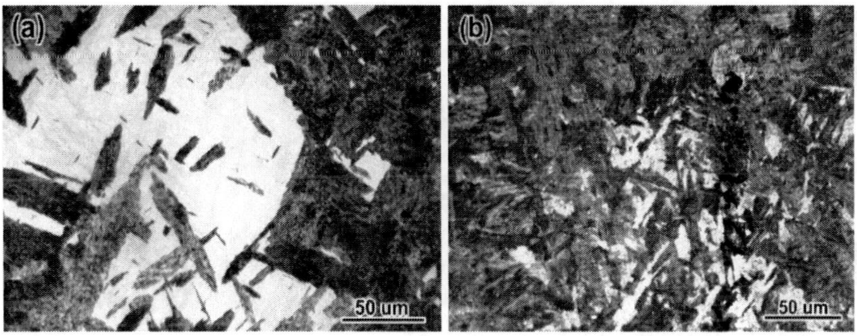

Figure 4.4. Microstructure of steel 75Mn. Holding at 350°C for: (a) 1 min, (b) 3 min.

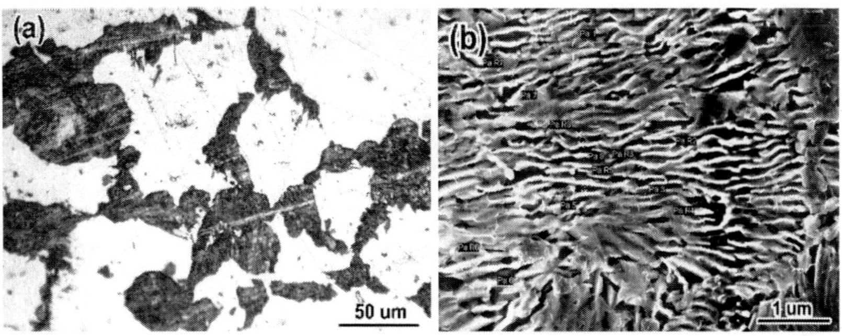

Figure 4.5. Microstructure of steel 75Mn. Holding at 400°C for: (a) 10 sec, (b) 10 min.

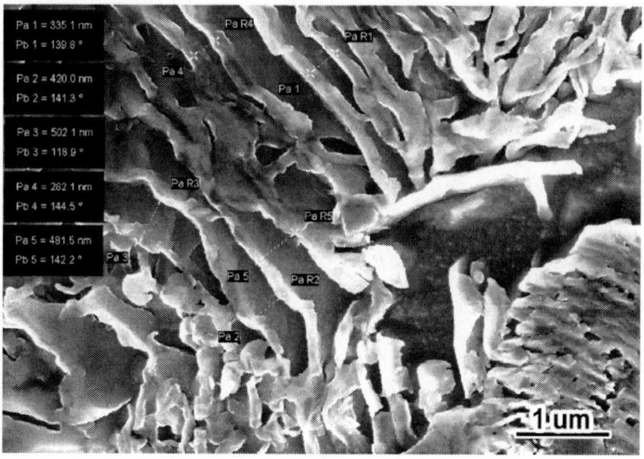

Figure 4.6. Microstructure of steel 75Mn. Holding at 600°C for 10 min.

The results reveal that bainitic transformation occurs in steel 75Mn under isothermal conditions at temperatures not higher than 350°C. Thus, for this steel, the lower boundary of the diffusional transformation lies somewhere below the "nose" of the TTT diagram. This confirms the data of [51] about formation of a low-temperature pearlite in high carbon steels at temperatures of 375-325°C.

The kinetic curves of the austenite transformation in 75Mn2Si steel under isothermal conditions are presented in Figure 4.7,a. As follows from Figure 4.7,b, steel 75Mn2Si is characterized by a C-shaped TTT diagram with the transformation kinetic maximum corresponding to 500-550°C. The "nose" of the diagram corresponds to 550°C (τ_{inc} =7 sec); at 450- 500°C the incubation period is only by 2-5 sec longer than that at 550°C. Martensitic point Ms (185°C) of steel 75Mn2Si is lower than that in steel 75Mn due to the higher content of manganese.

The transformation of overcooled austenite at 200°C begins with the formation of short (somewhat thickened) needle-like precipitates that occur predominantly along the grain boundaries (Figure 4.8,a). As the transformation develops, these "needles" do not thicken, but new (thinner) "needles" are formed mainly inside the grains along the crystallographic planes (Figure 4.8, b, c). The transformation stops while maintaining a significant proportion of austenite, which partially transforms into fine-needle ("structureless") martensite (Figure 4.8, d). The microhardness of the structure obtained at 200°C is about 700-710 HV.

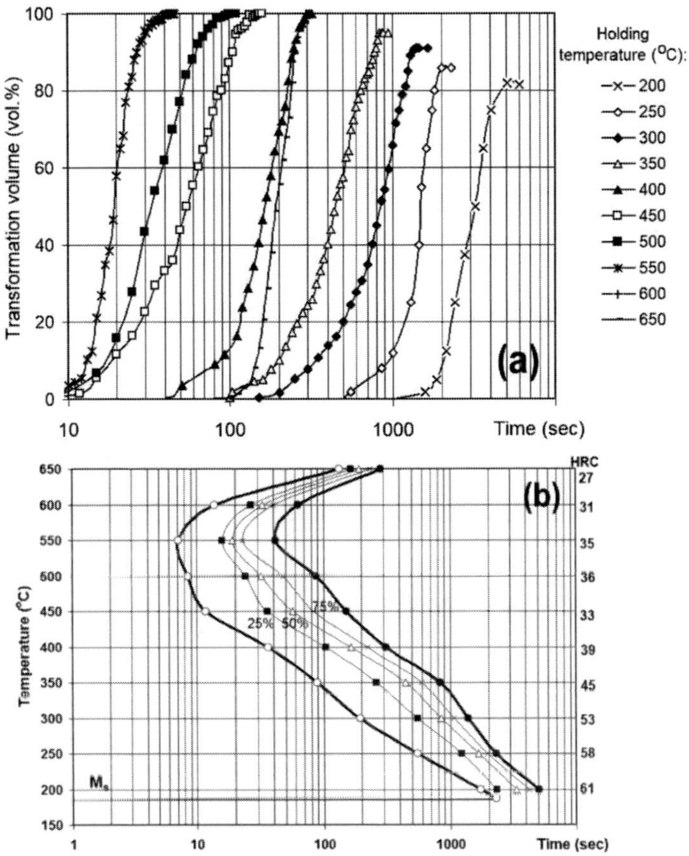

Figure 4.7. (a) Transformation kinetic curves and (b) TTT diagram for steel 75Mn2Si.

The transformation occurring at 250-300°C is characterized by the emergence of short thin "needles" in the first stage, which then enlarge and merge with newly formed "needles" into bulky dark areas (Figure 4.9, a). The transformation develops more active (as compared with 200°C), leaving fewer non-transformed areas. The microhardness of the steel transformed at 250-300°C is 578-661 HV. When going from 300°C to 350°C, the microhardness of steel decreases to 452 HV, while the shape of bainite transformation products changes from "needle-like" to plate (thickened), while maintaining an orientation along certain crystallographic planes (Figure 4.10, a). With an increase in holding temperature from 400°C to 450°C, the bainitic plates become thicker (relative to 350 °C), acquiring a heterogeneous "feathery" character (Figure 4.10, c). Such structure has a decreased micro-hardness (337 HV).

In the temperature range of isothermal holding of 500-650°C, pearlitic transformation occurs in steel 75Mn2Si, with the formation of troostite/sorbite structure at 500°C, sorbite structure at 550°C, and sorbite/coarse pearlite structure at 650°C (Figure 4.10, d). As the holding temperature increases, the microhardness of the transformation products decreases from 360 HV at 500°C to 250 HV at 650°C.

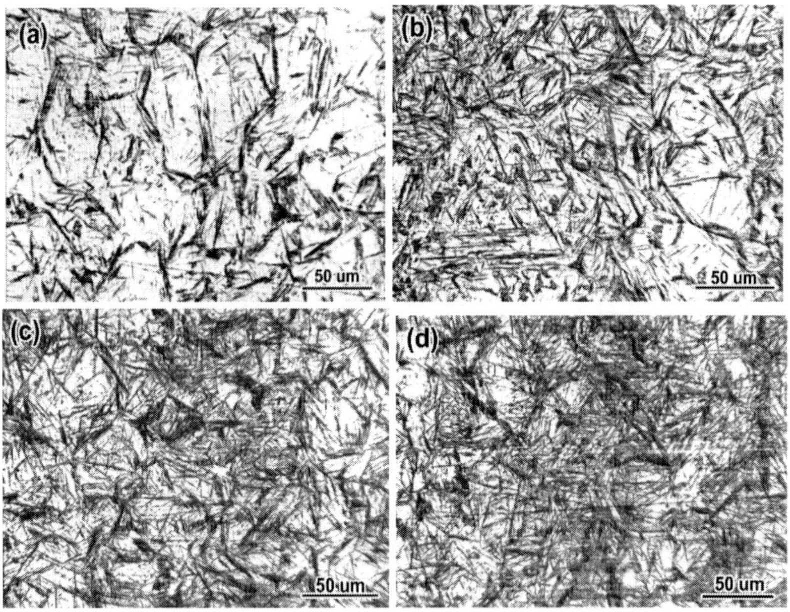

Figure 4.8. Microstructure of steel 75Mn2Si obtained after holding at 200°C for: (a) 2100 sec, (b) 3000 sec, (c) 4000 sec, (d) 5200 sec.

The peculiarities of austenite transformation in steel 75MnCrSiMo under isothermal conditions are presented in Figures 4.11. As follows from these figures, steel 75MnCrSiMo performs a more complicated behaviour of austenite transformation as compared with two previous steels. Due to alloying with strong carbide-forming elements (Cr, Mo), steel 75MnCrSiMo is distinguished by a TTT diagram *divided* into two separate *C-shape* domains. This means that the transformation of austenite in steel 75MnCrSiMo is characterized by two kinetic maxima: at 450°C (τ_{inc} = 50 sec) and 600°C (τ_{inc} = 38 sec), which is significantly higher than that in steels 75Mn and 75Mn2Si. The martensitic point of steel 75MnCrSiMo corresponds to 252°C.

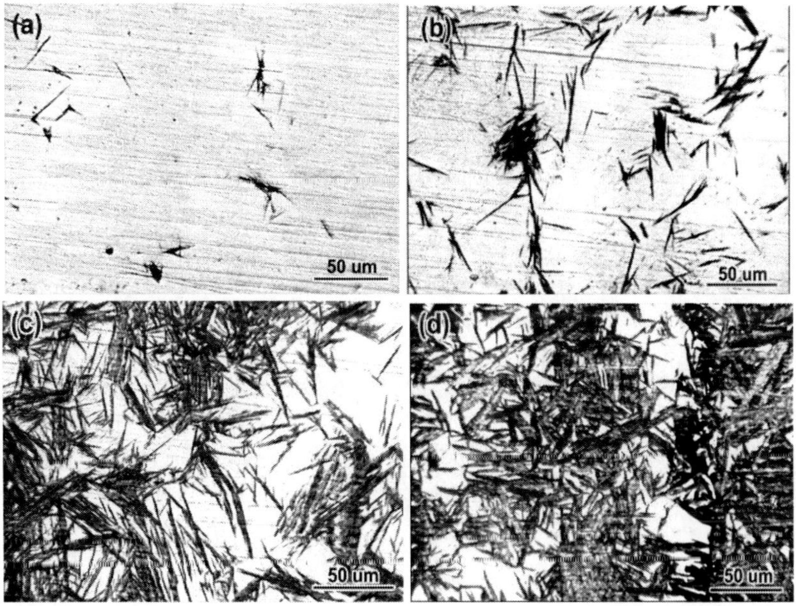

Figure 4.9. Microstructure of steel 75Mn2Si obtained after holding at 250°C for: (a) 800 sec, (b) 1200 sec, (c) 1700 sec, (d) 2500 sec.

Figures 4.12, 4.13 show the microstructure formed in steel 75MnCrSiMo at temperatures close to the Ms point. The products of bainite transformation at 200°C are long thin "needles" that do not thicken with transformation development. The transformation begins along the grain boundaries (where the enlarged needles appear) to proceed inside the intragranular regions (Figure 4.12, a). At 200°C, the transformation stops when a significant amount of austenite remains untransformed (Figure 4.12, d). A light background seen in micrographs indicates the presence of "structureless" martensite and retained austenite. The average microhardness of such a structure is about 683 HV.

Under holding at 250°C and 300°C the transformation proceeds with the formation of lens-like "needles", which are wider and look "massive" in comparison with t_{is}=200°C (Figures 4.13, 4.14). The transformation results in characteristic structure consisting of a "forest" of bainitic "needles", many of which are packed in packages of parallel "needles" (microhardness is 590-635 HV). Inside the packages, many "needles" are of lighter contrast, standing out against the background of other darker "needles".

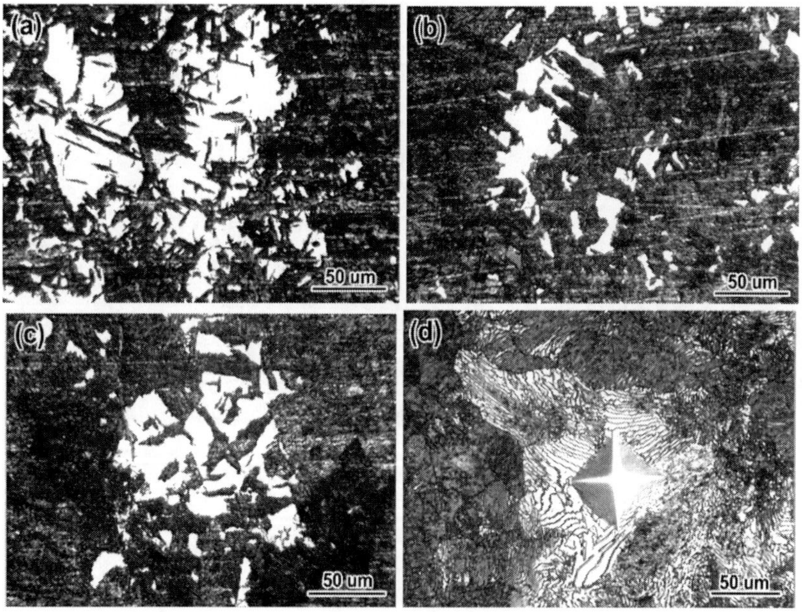

Figure 4.10. Microstructure of steel 75Mn2Si after completion of transformation at (a) 350°C, (b) 400°C, (c) 450°C, (d) 550°C.

With the developing transformation at 300°C, the "needles" thicken, merging with new "needles" to form massive conglomerates of darker contrast. The boundaries between individual needles are not visible inside the conglomerate, and the light areas have a more diffuse pattern (Figure

4.13, d). Comparison of Figures 4.13 with Figure 4.12 reveals that the structure formed at 250°C differs significantly from the structure obtained at a temperature below the Ms point: in the latter case, the needles are thinner and do not merge into bulky conglomerates.

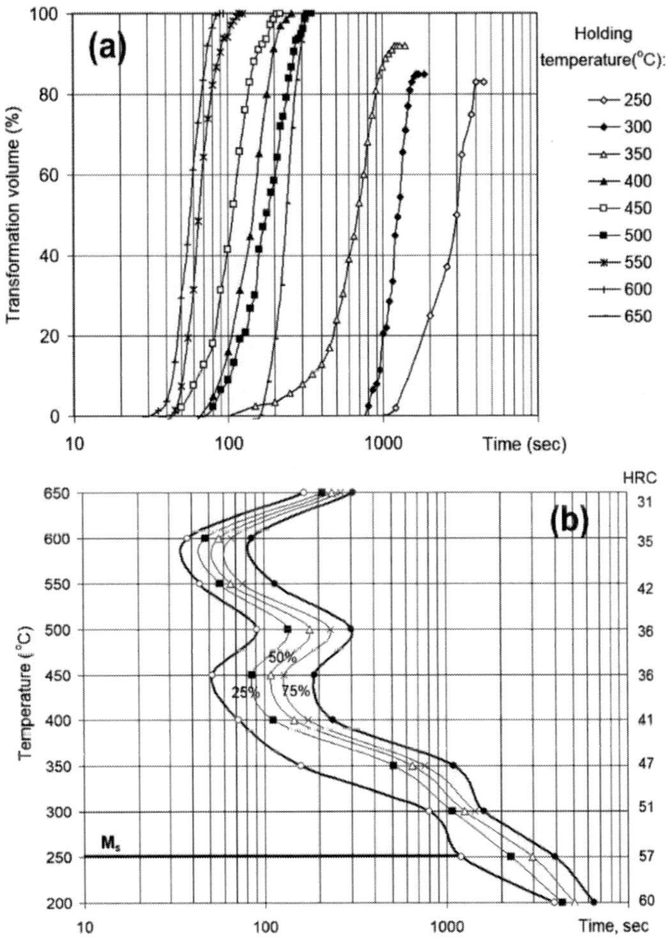

Figure 4.11. (a) Transformation kinetic curves and (b) TTT diagram for steel 75MnCrSiMo (From Kuz'min, S.O., Efremenko, V.G., Chabak, Yu. G., Tsvetkova, E.V. 2013. *Metallofiz. Noveishie Tekhnol.*, 35(9):1271–1282).

The bainitic structure, resulting from holding at 350-450°C (Figure 4.15), has a relatively uniform dark contrast, with none of the pronounced acicularization that is characteristic of low-temperature (200-300°C) bainite. At 350-400°C, the transformation begins with the appearance of elongated dark-contrast areas with different widths and jagged edges (Figure 4.15, a,c). In the course of the transformation, these areas grow quite quickly, resulting in grey-black "chequered" pattern (Figures 4.15, b,d). The structure formed at the temperature near the "nose" of the bainitic region (450°C) is distinguished by a clearly differentiated "feathery" pattern (Figure 4.15,e,f) with average microhardness of 354 HV.

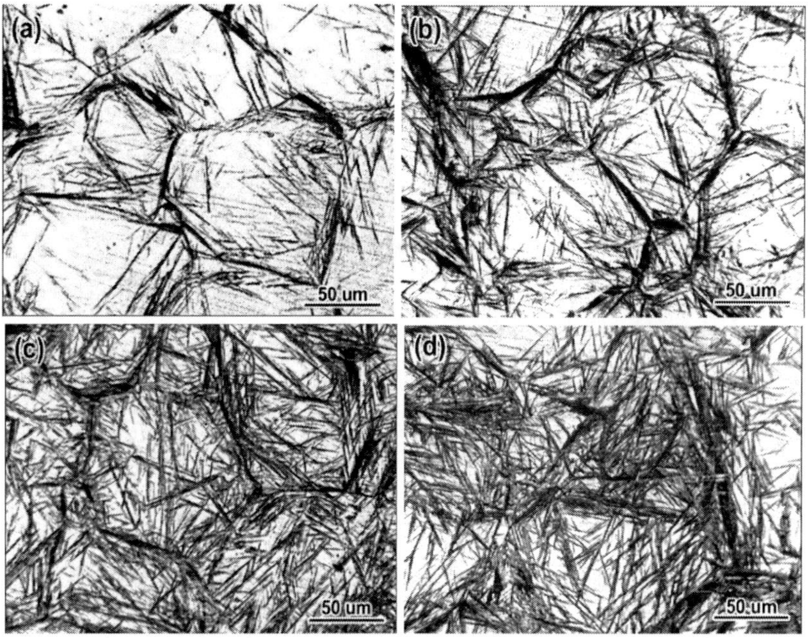

Figure 4.12. Microstructure of steel 75MnCrSiMo obtained after holding at 200°C for (a) 4150 sec, (b) 5000 sec, (c) 6000 sec, (d) 6500 sec (From Kuz'min, S.O., Efremenko, V.G., Chabak, Yu. G., Tsvetkova, E.V. 2013. *Metallofiz. Noveishie Tekhnol.*, 35(9):1271–1282).

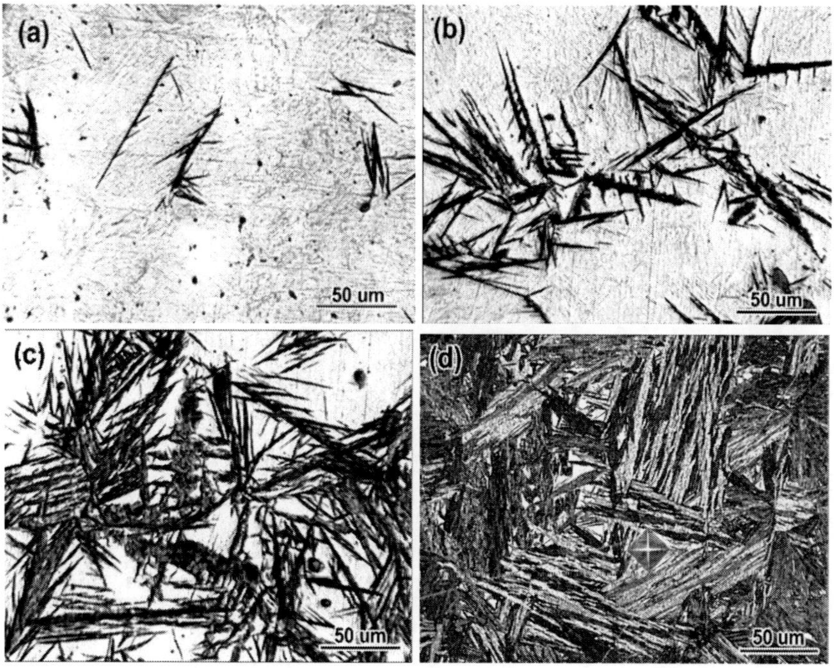

Figure 4.13. Microstructure of steel 75MnCrSiMo obtained after holding at 250°C for (a) 1300 sec, (b) 1800 sec, (c) 3000 sec, (d) 4500 sec (From Kuz'min, S.O., Efremenko, V.G., Chabak, Yu.G., Tsvetkova, E.V. 2013. *Metallofiz. Noveishie Tekhnol.*, 35(9):1271–1282).

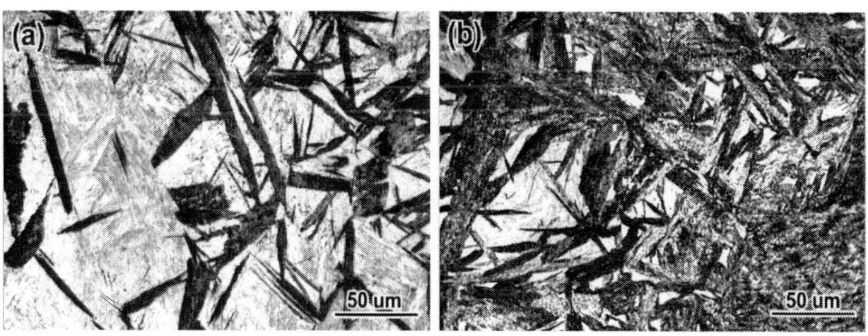

Figure 4.14. Microstructure of steel 75MnCrSiMo obtained after holding at 300°C for (a) 640 sec and (b) 1200 sec (From Kuz'min, S.O., Efremenko, V.G., Chabak, Yu. G., Tsvetkova, E.V. 2013. *Metallofiz. Noveishie Tekhnol.*, 35(9):1271–1282).

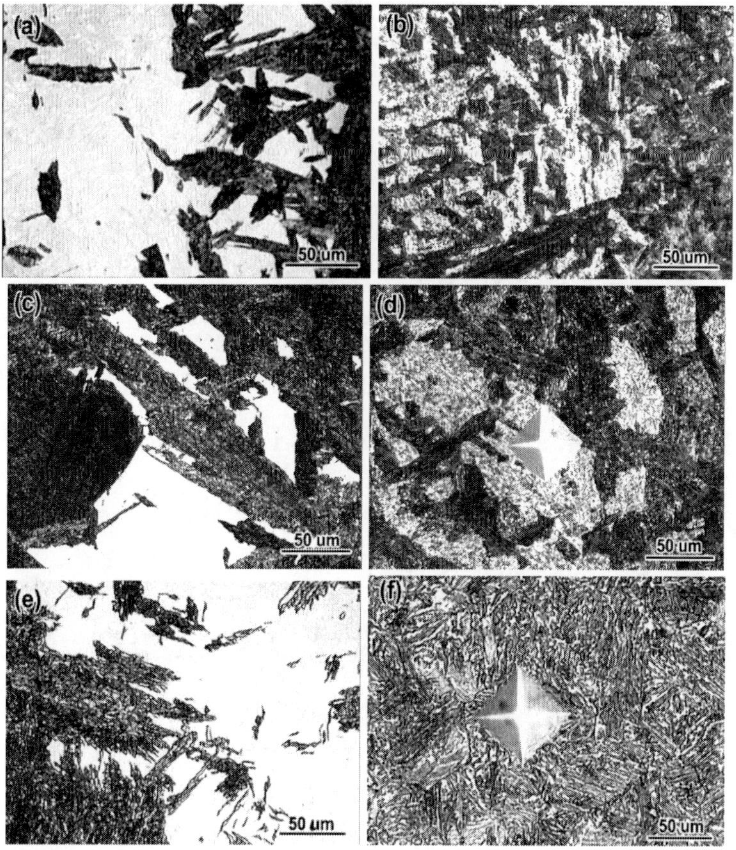

Figure 4.15. Microstructure of steel 75MnCrSiMo obtained after holding: at (a) 350°C for 3600 sec and (b) 6000 sec; at 400°C for (c) 120 sec and (d) 180 sec; at 450°C for (e) 60 sec and (f) 120 sec (From Kuz'min, S.O., Efremenko, V.G., Chabak, Yu.G., Tsvetkova, E.V. 2013. *Metallofiz. Noveishie Tekhnol.*, 35(9):1271–1282).

Austenite transformation at 500°C begins with the nucleation of small troostite colonies along the grain boundaries. In contrast to the rounded shape that is characteristic of troostite in steels 75Mn and 75Mn2Si, in this case the troostite colonies have an irregular shape with "torn" edges (Figure 4.16,a). The development of the eutectoid transformation is accompanied by the appearance of thin areas of upper bainite of white contrast (Figure 4.16, b). Such heterogeneous structure has a lower microhardness (356 HV on average).

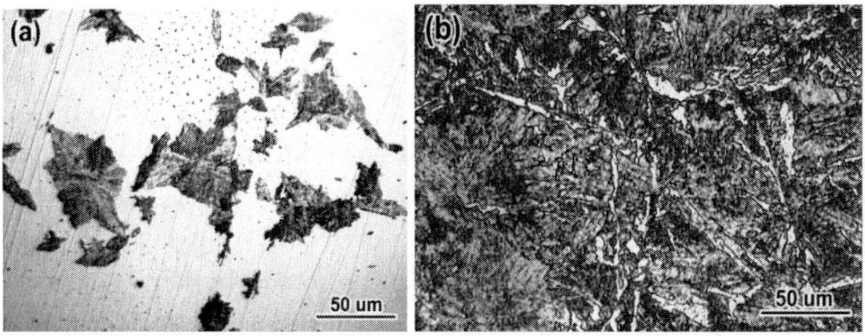

Figure 4.16. Microstructure of steel 75MnCrSiMo. Holding (a) at 500°C for 120 sec and (b) 300 sec.

Structure formation in steel 75MnCrSiMo under isothermal holding at 550-650°C proceeds via pearlite (reconstructive) mechanism, with the formation of troostite/sorbite structure at 550°C (416 HV) and sorbite structure at 600°C (343 HV) and 650°C (310 HV).

Data on the types and morphological features of the microstructure formed in the studied steels under isothermal conditions were used to interpret the nature of structure formation that occurs during continuous cooling.

5. Structure Formation Steels under Continuos Cooling

Samples of steel 75Mn heated to 850°C were then cooled at different rates in the range of 6.5-33.3 K/sec (the rates were taken as the average value for the temperature range of 800-200°C). The Continuous-Cooling-Temperature diagram (CCT) for steel 75Mn is presented in Figure 5.1.

As follows from Figure 5.1, steel 75Mn belongs to the steels with a pearlitic type of hardenability; its CCT indicates the presence of only pearlitic and martensitic transformation areas. At cooling rates in the range of 6.5-14.3 K/sec, the transformation takes place in the temperature range of 538-620°C, resulting in the formation of eutectoid structure (Figure 5.2, a). The microhardness of the structure increases from 335 HV to 382 HV as the cooling rate increases, having a different pearlite dispersion and, accordingly, different microhardness. At a cooling rate of 16.0 K/sec, the austenite transformation begins at 525°C with the formation of troostite with a microhardness of 444-465 HV, after which it stops and continues only after reaching the martensitic domain; as a result, a troostite/martensite structure is formed with a predominance (70- 75%) of troostite (Figure 5.2, b).

At a cooling rate of 28.6 K/sec, troostite appears at 470-480°C, while most of the austenite is being overcooled to the martensitic range. The resultant structure is martensite (795-815 HV) and a minor amount (5-10 vol.%) of troostite lying along the boundaries of the former austenitic grains (Figure 5.2, c). After cooling at a rate of 33.3 K/sec, the structure is fine-needle martensite with a hardness of 835-846 HV (Figure 5.2, d). Thus, the cooling rate lying between 28.6 K/sec and 33.3 K/sec could be considered as the critical cooling rate for steel 75Mn when quenched for martensitic structure.

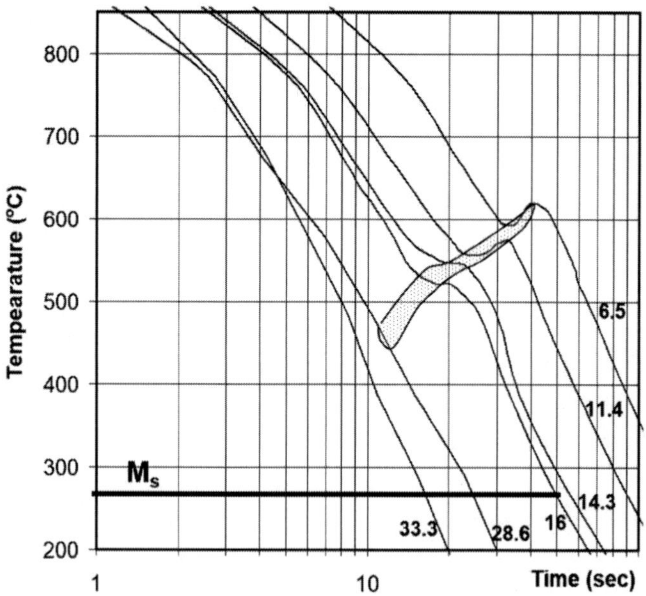

Figure 5.1. CCT diagram for steel 75Mn. The values at cooling curves are cooling rate (K/sec).

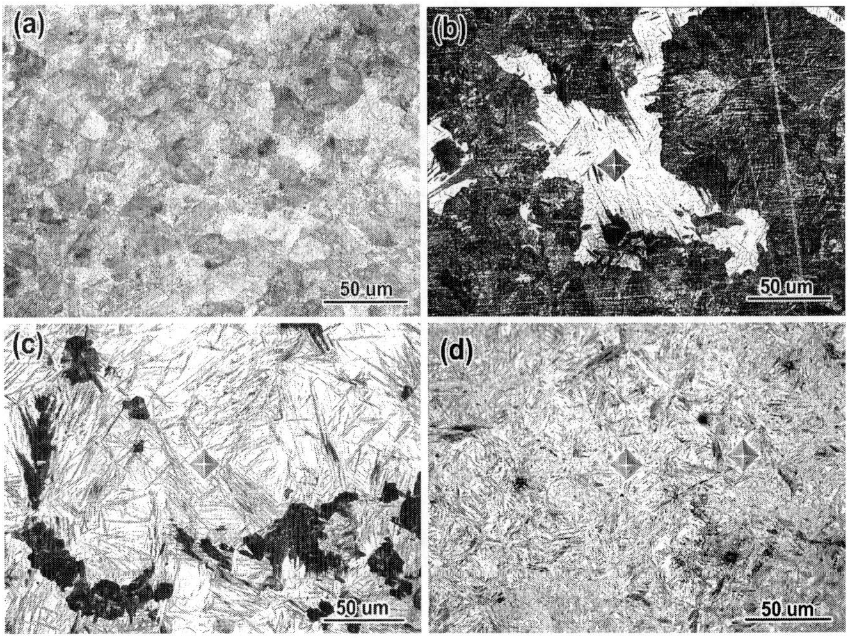

Figure 5.2. The microstructure of steel 75Mn cooled with cooling rate of (a) 6.5 K/sec, (b) 16 K/sec, (c) 28.6 K/sec, (d) 33.3 K/sec.

CCT diagram for steel 75Mn2Si is depicted in Figure 5.3. Analyzing these curves showed that an increase in manganese content by 0.45 wt.% (to 1.40 wt.%) and silicon by 0.1 wt.% (relative to steel 75Mn) resulted in a significant increase in the stability of the overcooled austenite in the pearlitic and martensitic temperature ranges. The pearlite transformation domain shifted to the right along the time scale (τ_{inc} increased from 11 to 33 sec) while the bainitic transformation domain appeared at 400-490°C, and the martensitic point Ms decreased by 75°C (from 260°C to 185°C).

When cooled at a rate of 14.8 K/sec, a martensitic structure with a hardness of 768-825 HV appeared (Figure 5.4, a). With a decrease in the cooling rate to 10.2 K/sec, occasional elongated areas of the upper bainite (which formed at 450-400°C) are observed along with martensite (809-817 HV) (Figure 5.4, b). Thus, steel 75Mn2Si refers to the steels with the bainitic type of hardenability. Since at a cooling rate of 10.2 K/sec the amount of bainite does not exceed 5 vol.%, it can be assumed that this is the critical cooling rate for steel 75Mn2Si when quenched for a martensitic structure. At a cooling rate of 8.5 K/sec, a multiphase structure forms, consisting of martensite (major), upper bainite and troostite (Figure 5.4, c). At a lower cooling rate (5.6 K/sec), the amount of martensite decreases sharply and is replaced by a pearlite formed in the range of 555-565°C (Figure 5.4, d). At cooling rates from 5.6 K/sec and below, the transformation proceeds completely in the pearlitic region at temperatures of 580-650°C.

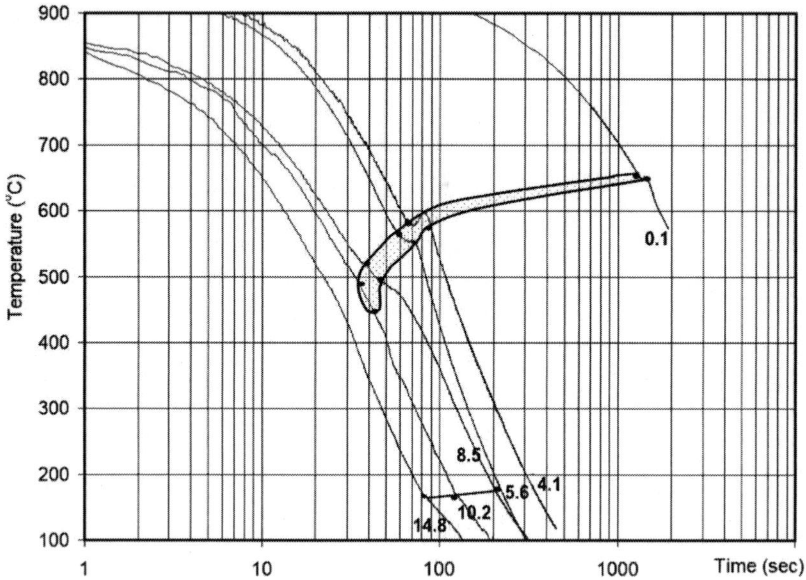

Figure 5.3. CCT diagram for steel 75Mn2Si. The values at curves are cooling rate (K/sec).

The CCT diagram for steel 75MnCrSiMo is shown in Figure 5.5. According to this diagram, when cooled at a rate of 36.7 K/sec, the transformation of austenite begins at 252°C (i.e., when reaching the Ms point), and a martensite structure with a hardness of 825-836 HV is formed. Reducing the cooling rate to 13.2 K/sec leads to the formation at 345-355°C of a minor (less than 10 vol.%) amount of upper bainite as separate wide plates (Figure 5.6,a). At this cooling rate, the major part of the austenite transforms into martensite (804-825 HV) at temperatures below 255°C.

Thus, the cooling rate of 13.2 K/sec is close to the critical rate for steel 75MnCrSiMo when quenched for martensitic structure.

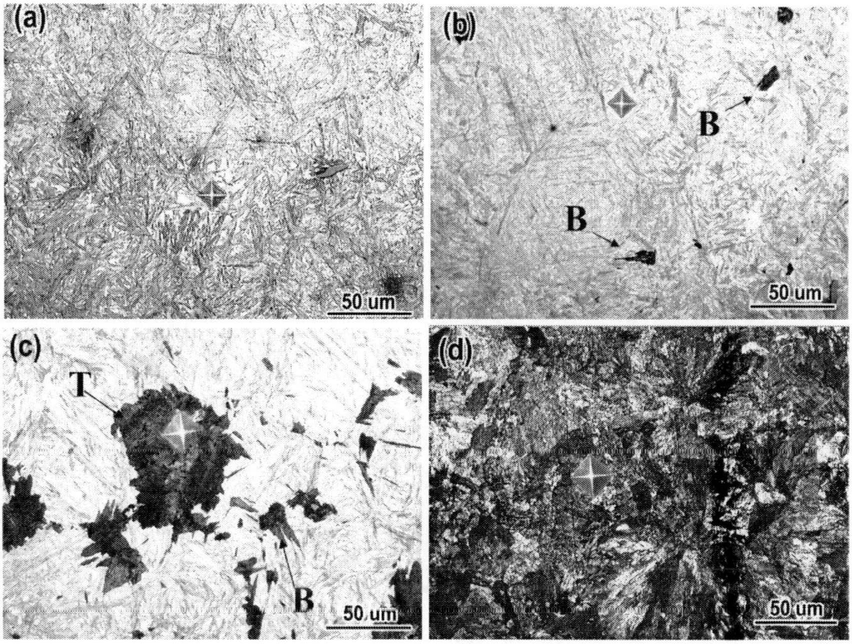

Figure 5.4. The microstructure of steel 75Mn2Si cooled with cooling rate of (a) 14.8 K/sec, (b)10.2 K/sec, (c) 8.5 K/sec, (d) 4.1 K/sec.

With further decrease in the cooling rate to 3.0 K/sec, the temperature range of the bainite transformation increases to 440-385°C, while the martensitic point Ms decreases to 230°C. The decrease in the cooling rate is accompanied with an increase in the volume fracture of bainite in the structure (Figure 5.6,b). The microhardness of the bainitic and martensitic components varies within 404-596 HV and 764-837 HV accordingly. At a cooling rate of 0.85 K/sec, the transformation begins in the pearlitic domain at 570-595°C and then proceeds in the bainitic domain below 460°C. As a result, a uniform dark structure with a hardness of 371-405 HV is formed, as shown in Figure 5.6,c.

Decreasing the cooling rate to 0.63 K/sec or less increases the range of pearlite transformation to 630-670°C. In this case, a eutectoid structure of different dispersity is formed in the steel. Thus, at a cooling rate of 0.65 K/sec, a dispersed troostite/sorbite structure with a hardness of 370-391 HV is formed, while at a cooling rate of 0.14-0.20 K/sec, a sorbite/pearlite structure with a hardness of 326-364 HV appears (Figure 5.6, d). After cooling at a rate of 0.03 K/sec, the amount of coarse pearlite increases, and the hardness decreases to 304-321 HV.

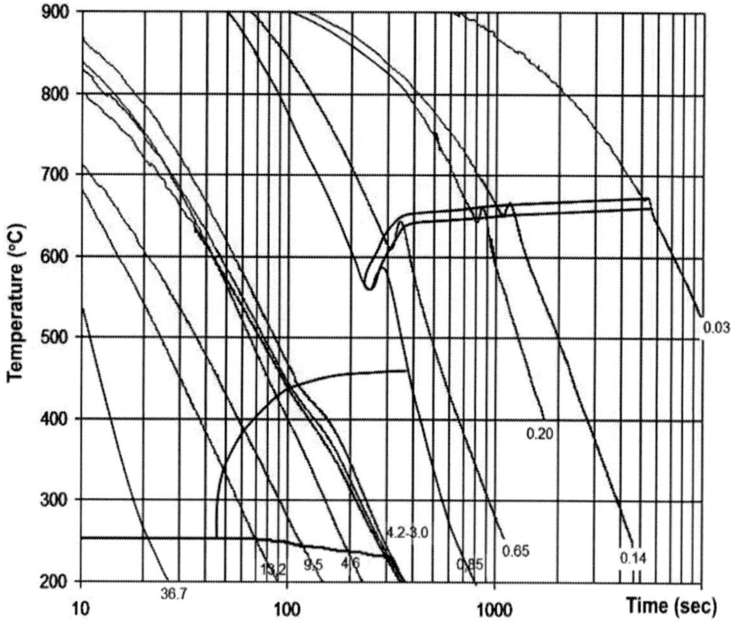

Figure 5.5. CCT diagram for steel 75MnCrSiMo. The values at curves are cooling rate (K/sec).

Summing up, steel 75MnCrSiMo refers to steels with a bainitic type of hardenability, i.e., in it, with continuous cooling, the region of the pearlite transformation is shifted to the right relative to the interval of bainite transformation. The cooling rate of 13.2 K/sec can be considered as the critical rate for this steel when quenched for the martensitic structure. In the range of cooling rates of 13.2-3.6 K/sec, the transformation begins in the bainite area and finishes in the martensitic area; the decrease in the cooling rate is accompanied by a slight decrease in the martensite microhardness. At the cooling rate of 0.85 K/sec, the transformation of austenite begins with the formation of pearlite and finishes with the formation of bainite. At cooling rates less than 0.65 K/sec, the whole overcooled austenite transforms into a eutectoid of various interlamellar distances.

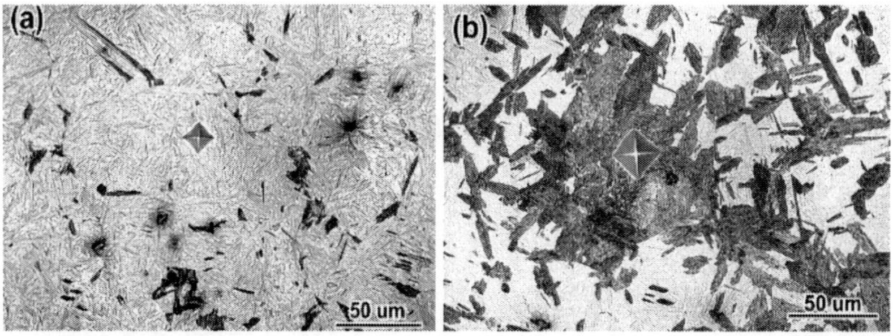

Figure 5.6. (Continued)

Figure 5.6. The microstructure of steel 75MnCrSiMo cooled with cooling rate of (a) 13.2 K/sec, (b)3.0 K/sec, (c) 0.85 K/sec, (d) 0.2 K/sec.

6. Comparative Analysis of the Morphology of Austenite Transformation Products

The data presented above allow us to conclude that the studied low-alloyed steels, which have almost the same carbon content (about 0.75 wt.%), differ in the morphology of the structures formed in the bainitic temperature domain. The generalized data on the temperature ranges for the formation of various types of bainite structure are presented in Figure 6.1.

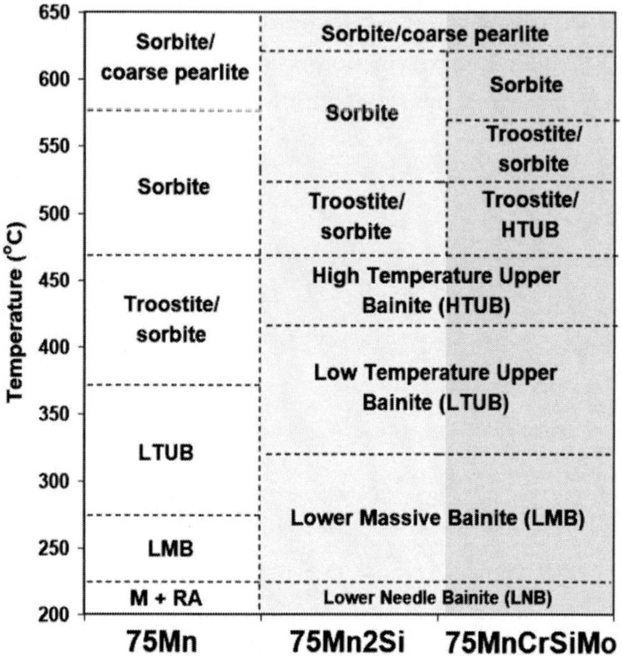

Figure 6.1. Temperature ranges for the formation of various types of microstructure in the studied steels under isothermal holding.

During isothermal holding at 200°C, the lower bainite of fine-needle morphology is formed in steels 75Mn2Si and 75MnCrSiMo. This kind of structure was not noted for steel 75Mn held at 200°C.

At temperatures in the range of 250-300°C, the transformation in the studied steels also proceeds through the formation of needle-shaped (lens-shaped) laths. However, as needle-like laths are formed, further they turn to gain "massive" morphology because of the thickening and merging of the "needles" into packages. This results in lower temperature "massive" bainite formation. This kind of bainite is detected in steel 75Mn in the fairly narrow temperature range of 225-250°C, while in steels 75Mn2Si and 75MnCrSiMo the temperature range is somewhat wider (250-300°C). It should be noted that at lower holding temperatures the first portions of bainite appear in steel 75Mn inside the austenite grains, while in steels 75Mn2Si and 75MnCrSiMo they are mainly along the grain boundaries. Presumably, this is because the bulk diffusion of carbon at low temperatures in steels 75Mn2Si and 75MnCrSiMo is more difficult than that in steel 75Mn, so the transformation is realized mainly through grain-boundary diffusion, developing mostly by increasing the amount of acicular laths (not by their thickening). Taking into account the morphological features (needle-like structure), the products of bainite transformation revealed in the studied steels after holding at 200-300°C are suggested to be considered as the lower bainite of two subgroups, namely "Lower Needle Bainite" (LNB), observed in steels 75Mn2Si and 75MnCrSiMo, and "Lower Massive Bainite" (LMB), which is observed in all three steels.

When the transformation temperature rises to 350°C, the bainite shape changes from needle-like to laths-like, which can be interpreted as a transition from lower bainite to upper bainite. The formation of upper bainite begins with the appearance of occasional small thickened plates (mainly along the grain boundaries). As the transformation progresses, these plates subsequently grow in the lateral and forward directions. The transformation is completed by the formation of either a homogeneous (in contrast) dark structure or a characteristic "chequered" structure with alternating light-grey and dark-grey areas oriented along different crystallographic directions.

In steel 75Mn, the upper bainite is formed at 300-350°C where it has the same morphology for this temperature range. In steels 75Mn2Si and 75MnCrSiMo, the temperature interval is wider (350-450°C), which leads to differences in the morphology of the upper bainite. In its lower part (350-400°C), the transformation follows the scheme described above and ends with the formation of a structure similar to the upper bainite in steel 75Mn. At 450°C, a pronounced heterogeneous (feathery) microstructure is formed in steels 75Mn2Si and 75MnCrSiMo, when the heterogeneity of the structure is seen from the very beginning of transformation (Figure 4.15,e). Thus, the temperature range for upper bainite structure in steels 75Mn2Si and 75MnCrSiMo may be divided into two ranges referring to different structural modifications, which are proposed to be designated as "Low Temperature Upper Bainite" (LTUB) and "High Temperature Upper Bainite" (HTUB).

The proposed division of the products of bainitic transformation into subgroups and the new terms (LNB, LMB, LTUB, HTUB) are introduced for the first time to be discussed; these terms make sense only in relation to the studied steel grades. An analysis of literary sources [19–21] shows that such an approach to describing bainitic structures is already widely used in practice, since the diversity of the morphology of bainitic transformation products in various steels often does not fit the traditional division of structures into “upper” and “lower” bainite. This has led researchers to suggest their own classifications which may be more appropriate for specific cases.

The upper bainite observed in steel 75Mn, due to its morphology, can be referred to LTUB. The data obtained show that in steel 75Mn the transformation develops through the pearlite mechanism at a rather low holding temperature (400°C), i.e., below the “nose” of the C-shaped TTT diagram. In steels 75Mn2Si and 75MnCrSiMo, the transition from the bainitic mechanism to the pearlitic one is shifted to 500°C. In steels 75Mn2Si and 75MnCrSiMo, the transition from the bainitic mechanism to pearlitic mechanism is shifted to 500°C when some portion of HTUB structure can be detected by the acicular ferrite component. Thus, at 500°C, the steel 75MnCrSiMo transforms with an overlapping of diffusional (reconstructive) and bainite transformation mechanisms.

Approximately the same types of microstructure are observed in all three steels in the temperature range of the pearlite transformation. However, it should be noted that at the same temperature of transformation, the eutectoid formed in steel 75MnCrSiMo has a more dispersed structure and, accordingly, a higher (by 25-50 HV) microhardness as compared to steels 75Mn and 75Mn2Si.

The results of XRD measurements of all the studied steels subjected to isothermal holding are presented in Figure 6.2 (the holding duration ensured the completion of phase transformation at a certain temperature). As follows from this figure, the higher amount of retained austenite (12.5-14 vol.%) is attributed to bainite in steel 75Mn2Si. The volume fraction of retained austenite in bainite of steel 75MnCrSiMo is 6.2-7.5 vol.%. In steel 75Mn, isothermally treated at 250°C, retained austenite was not found. The presence of 11 vol.% of retained austenite in a steel treated at 200°C (below the martensitic point) is explained by the thermal stabilization of austenite since bainitic transformation does not proceed at 200°C in steel 75Mn with long holding (up to 10 thousand sec).

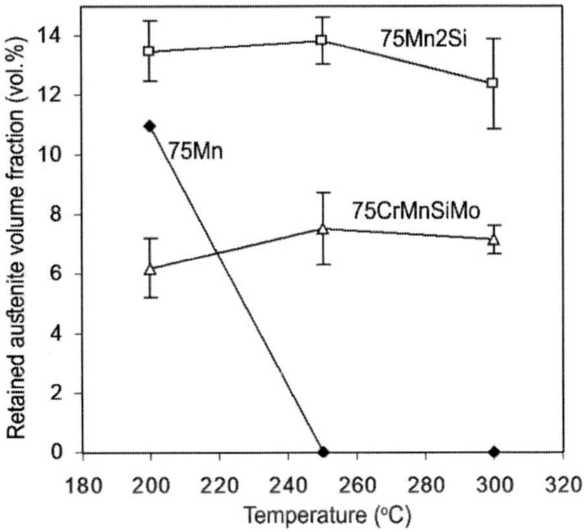

Figure 6.2. The effect of isothermal holding temperature on the amount of retained austenite in steels 75Mn, 75Mn2Si and 75MnCrSiMo.

From the scientific point of view, the morphology of the structures forming in the lower temperature range is of great interest. Transmission electron microscopy was used to clarify the features of the fine structure of bainite obtained in steels 75Mn, 75Mn2Si and 75MnCrSiMo by isothermal holding in the temperature range of 200-450°C. The durations of holding corresponded to the completion of the transformation were chosen according to the corresponding TTT diagrams. Using the TEM images obtained, the common character of the structure was evaluated and the quantitative parameters of the microstructural components were determined, specifically the average values of the following parameters: a) the width of the ferrite laths; b) the length and width of carbide particles; c) carbide volume fraction and density distribution (number of particles per square micrometer (pcs/μ^2)) (Figure 6.3).

TEM observation revealed that LMB (250°C) in steel 75Mn consists of α-phase laths of 0.5-1.0 μm width (Figure 6.4, a). Inside these laths, carbides lie at an angle about 60 degrees to the long axis with increased density distribution (12.6 pcs/μm²). Carbide average length and width are 0.15 μm and 0.022 μm, respectively. The boundaries of the laths are free of carbides. The ferrite laths have a higher dislocation density seen as dislocations "coils" and clusters (Figure 6.4, b).

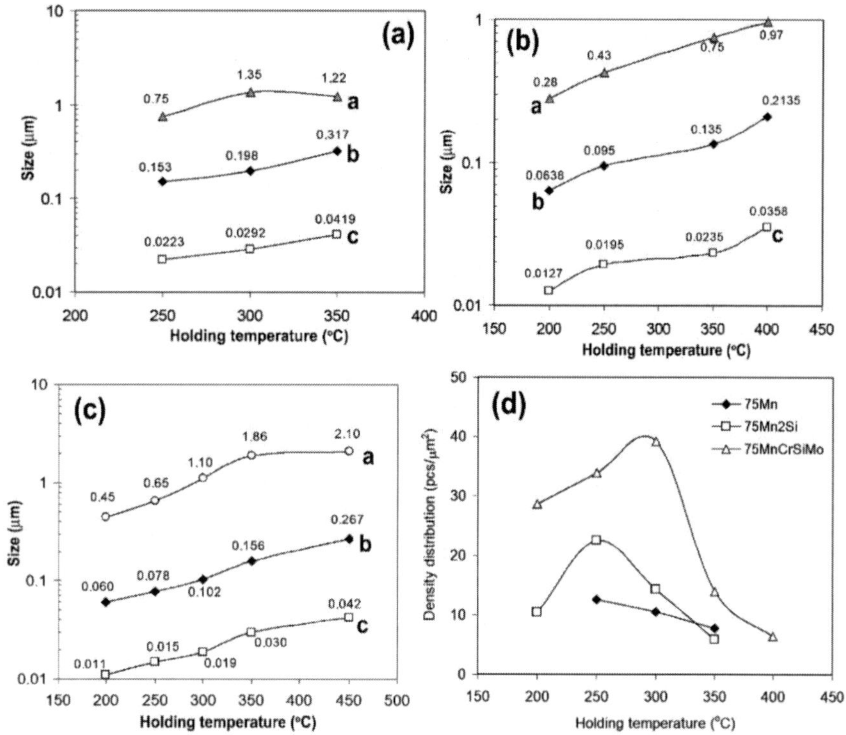

Figure 6.3. The effect of isothermal holding temperature on the sizes of carbides and ferrite laths in the steels (a) 75Mn, (b) 75Mn2Si, and (c) 75MnCrSiMo (at the curves: a – carbide length, b – carbide width, c - ferrite laths width). (d) Density distribution of carbides as a function of holding temperature. (From Kuz'min, S.O., Efremenko, V.G., Chabak, Yu. G., Tsvetkova, E.V. 2013. *Metallofiz. Noveishie Tekhnol.*, 35(9): 1271–1282).

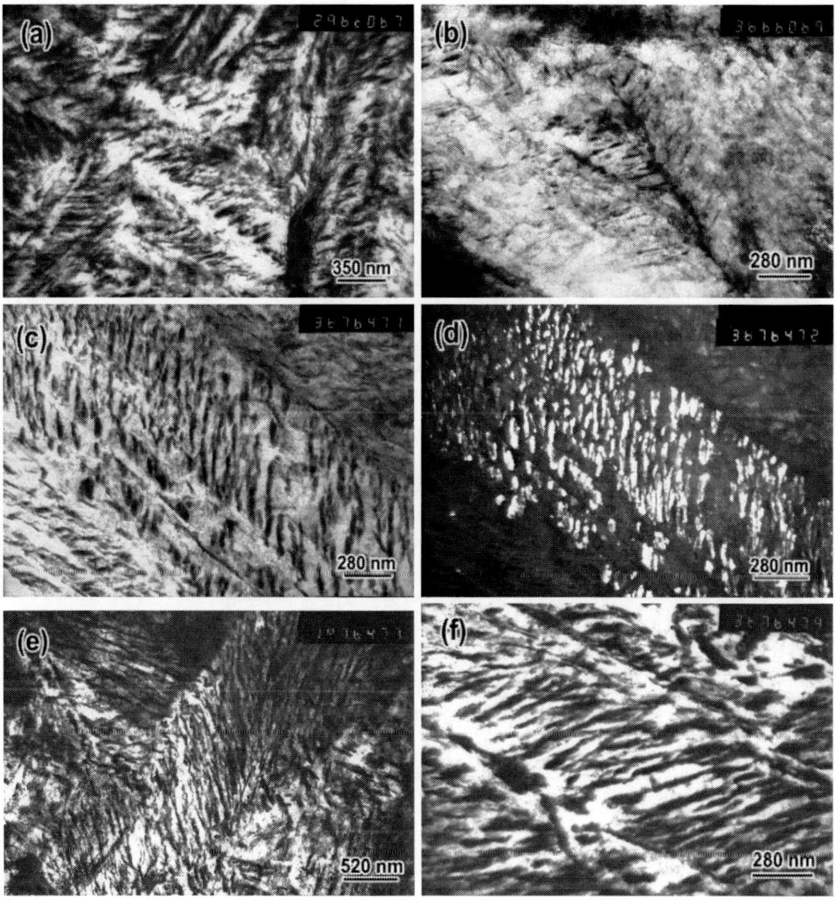

Figure 6.4. TEM images of microstructure of steel 75Mn after isothermal holding at: (a, b) 250°C, (c, d) 300°C, (e, f) 350°C ((d) dark-field image of (c) in reflection of cementite).

Within the laths, the mosaic blocks are disoriented, resulting in blurring of α-phase reflections on the SAED (Selected Area Electron Diffraction) images. An increase in the transformation temperature to 300°C led to the formation of LTUB in 75Mn steel. The average width of the α-phase laths is increased to 1.0-1.8 μm. The carbide particles are enlarged: the average carbide length and width increased by about 30% compared to holding at 250°C (Figure 6.4, c,d). The coarsening of carbides is accompanied by decrease in the density of carbides distribution (to 10.6 pcs/μm^2). The boundaries of the laths remain free of carbide inclusions and have a brighter contrast, which indicates a reduced density of lattice defects in the boundary areas due to the dislocation flow to the boundaries (Figure 6.4, c). Inside the laths, fragments located along the main axis of the lath are observed; the boundaries between the fragments are heterogeneous: there are both areas free from dislocations and areas formed by dislocation walls. Dark field images made in cementite reflections confirm carbides precipitation inside the laths at an angle of 50-57 degrees to the long axis; no carbides are observed along the laths boundaries (Figure 6.4, d).

LTUB, obtained at 350°C, is characterized by the presence of ferrite laths with an average width of 0.9-1.3 μm. Cementite particles are revealed in the SAED images; their average maximum size is 0.32 μm length and 0.042 μm width. The coarsening of carbides was accompanied by a decrease in the density of their distribution (to 7.8 pcs/μm^2) (Figure 6.4, e, f). The predominant occurrence of carbides inside α-phase laths was still observed, but the angle of carbides inclination with respect to the long axis of the laths varied within fairly wide limits, from 42 to 55 degrees. That indicates fragmentation of the lath with rotation of separate fragments relative to each other. The boundaries mostly remained free of carbides, however in some places carbide films between α-phase laths were observed (Figure 6.4, f).

As shown above, holding at 200°C results in formation of LNB in steel 75Mn2Si. At the end of the bainitic transformation, a significant amount of austenite is retained in the structure, being partially transformed into martensite when cooled to room temperature. For this reason, the micrographs of samples held at 200°C showed the presence of martensite plates surrounded by areas of retained austenite (Figure 6.5,a). Relaxation of the stresses arising during the lattice shear at the transformation occurred partly through a twinning mechanism, and partly through a dislocation slip mechanism. This is evidenced by the presence of twins and dislocation clusters within α-phase laths. In α-phase laths of 0.2-0.4 μm width the initial stage of carbide precipitation was recorded at an angle from 55 to 70 degrees with respect to the longitudinal axis of the lath. These carbides are very fine and their average length is 0.064 μm and width is 0.013 μm. The dislocations clots form the sub-boundaries inside ferrite laths. The α-phase laths are separated by the films of retained austenite.

LMB structure forms in steel 75Mn2Si under holding at 250°C consisting of ferrite laths of 0.3-0.5 μm width. The numerous dispersed carbides are seen inside the laths with increased distribution density reaching 22.6 pcs/μm^2 (Figure 6.5, b,c). Being aligned in chains, the carbides form a kind of elongated fibre embedded into the laths matrix. Between the laths, austenite interlayers are positioned. The boundaries of ferrite laths are free of carbide precipitates.

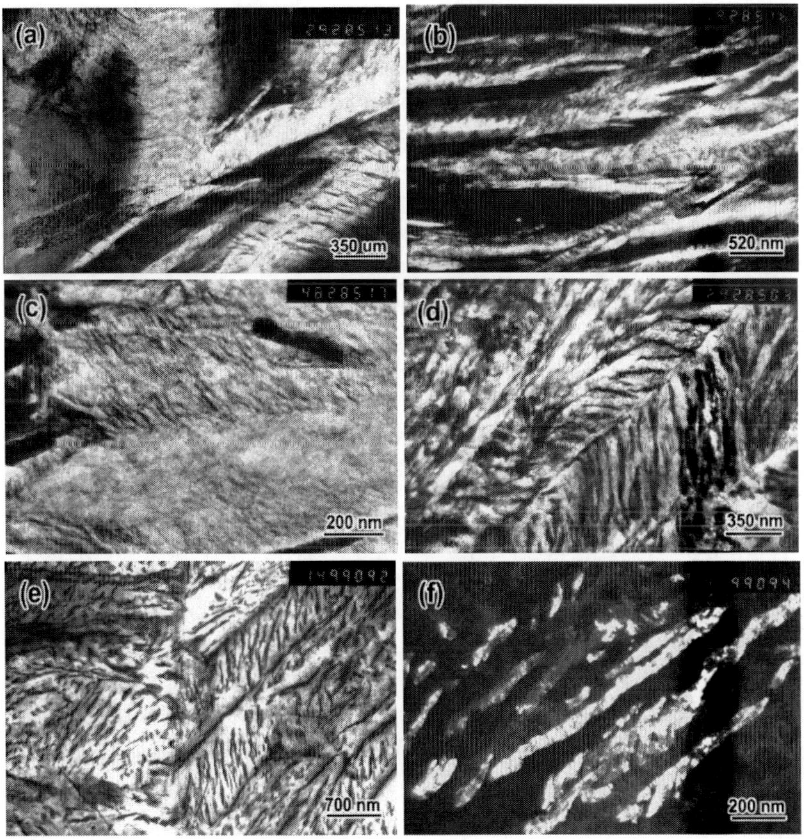

Figure 6.5. TEM images of microstructure of steel 75Mn2Si after isothermal holding at: (a) 200°C, (b, c) 250°C, (d) 350°C, (e, f) 400°C ((f) dark-field image in reflex of cementite).

Isothermal holding of steel 75Mn2Si at 350-400°C results in LTUB structure formation. The width of the laths in LTUB was estimated as 0.5-1.2 μm. LTUB, formed at 350°C, is characterized by higher carbide distribution density (14.4 pcs/μm^2) than that of LMB (Figure 6.5, d). The carbide particles are enlarged (up to 0.135 μm in length and up to 0.0235 μm in width). Many carbides are located inside the ferrite laths; enlarged carbides cross the laths almost from border to border. In some places elongated carbide inclusions are seen between the laths.

The distribution density of carbide particles in LTUB (400°C) is decreased to 5.92 pcs/μm^2 while carbide size reached maximum values: 0.21 μm in length and 0.04 μm in width. The enlargement of carbides led to an increase in their defectiveness allowing to observe carbide dislocation substructure. Carbides were still predominantly seen at an angle close to 60 degrees to lath long axis, but cementite carbides were also noted along the lath boundaries (see dark-field image in cementite reflection in Figure 6.5, e). In some areas ferrite laths divided by retained austenite films are seen.

The TEM images of bainite in steel 75MnCrSiMo are shown in Figure 6.6. LNB (200°C) consists of α-phase laths of 0.15-0.8 μm width intersecting at different angles; between them martensite and retained austenite are located (Figure 6.6, a). Within the laths, dispersed carbide precipitates (average length of 0.070 μm, width of 0.011 μm) are revealed in increased amounts (28.6 pcs/μm^2). The precipitates are oriented along the crystallographic planes at angles of ~56-70 degrees to lath long axis. The substructure of the laths is not uniform: micro-twins and dislocation clusters are observed along with areas that are relatively free of lattice defects (Figure 6.6, b).

After holding at 250°C, enlargement of the carbide particles inside the α-phase laths is noted with increase in the size and amount of carbide. Carbide-free boundaries are observed in some places, while in other places boundary-located carbide precipitations are seen (Figure 6.6, c). The specific feature of the laths substructure is that the laths consist of the fragments that are somewhat disoriented relative to each other; this is evidenced by the different crystallographic systems of carbide precipitation in adjacent fragments (Figure 6.6, d). As well, copious highly dispersed carbides were detected inside retained austenite areas.

In samples held at 300°C, a further increase in carbides distribution density is noted with an increase in carbides sizes. The carbides are mostly located inside the laths the width of which varies in the range of 0.75-1.75 μm with an average value of 1.10 μm. Meanwhile intermittent carbide precipitations are also detected along the laths boundaries (Figure 6.6, d). In general, the sample held at 300°C have coarser, more differentiated structure as compared with bainite formed at 250°C.

The formation of bainite at 350°C was accompanied by the precipitation of a lesser number of carbides but with a significant (two-fold) increase in their sizes (Figure 6.6, f,g). The microstructure is generally changed, having wide (2.5-4.0 μm) laths divided by interlaths elongated carbides. Between the wide laths, narrower laths (0.45-0.60 μm) are observed, with carbides at an angle of ~55 degrees to the long axis. The substructure of the laths contains the conglomerations of dislocations that are unevenly distributed between carbide particles.

The holding at 450°C led to HTUB formation with further coarsening of the structure (Figure 6.6, h). Compared with LTUB (350°C), the average width of the α-phase laths increased to 2.1 μm while the laths contained a lower density of lattice defects. The carbide size is increased on average by 1.5 times, while its density distribution reached its minimal value (6.5 pcs/μm^2). The carbides are mainly located along the ferrite laths. In some areas, the bending of carbides is observed, which indicates the rotation of laths fragments. The pronounced texturing of carbides gives a clear diffraction pattern in the SAED images.

TEM investigation revealed that the formation of bainite in steels 75Mn, 75Mn2Si and 75MnCrSiMo mostly corresponds to the generally recognized concept of bainite transformation mechanism. At the lowest holding temperature (200°C), "Lower Needle Bainite" appeared, consisting of α-phase laths with the carbides precipitated inside the laths at an angle close to 60 degrees to the laths long axis. The boundaries between the laths remain free of carbides. LNB contains an increased volume fraction retained austenite in the structure.

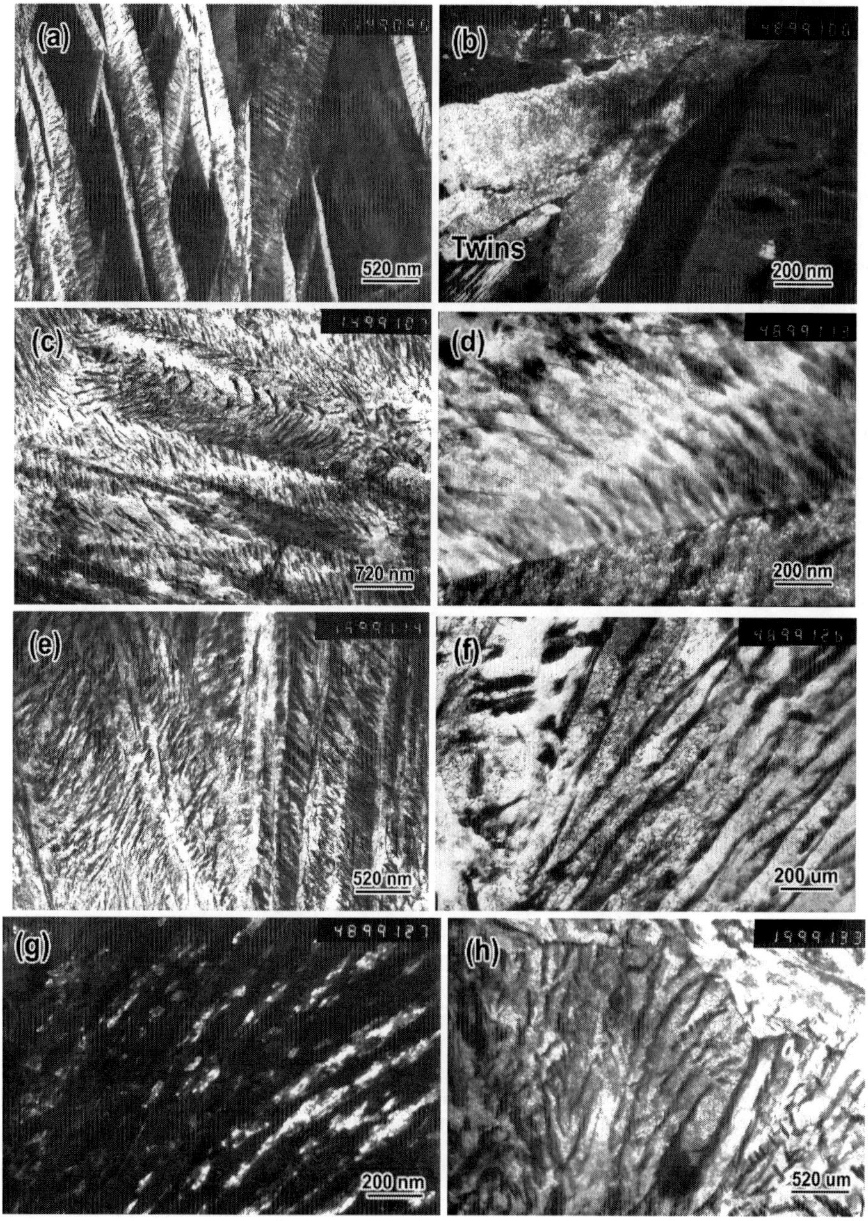

Figure 6.6. TEM images of steel 75MnCrSiMo after isothermal holding at: (a, b) 200°C, (c, d) 250°C, (e) 300°C, (f, g) 350°C, (h) 450°C((g) dark-field image of (f) in reflex of cementite) (From Kuz'min, S.O., Efremenko, V.G., Chabak, Yu. G., Tsvetkova, E.V. 2013. *Metallofiz. Noveishie Tekhnol.*, 35(9):1271–1282).

With an increase in the temperature of isothermal holding up to 300°C (in the steel 75MnCrSiMo - up to 250°C), carbides begin to precipitate not only inside the α-phase laths, but also along the laths boundaries. With the transition from LNB to the LTUB, a pronounced coarsening of the structure is observed. The carbides are still predominantly located at an angle of 60 degrees to the long axis, but the total number of carbides decreases, while they become larger and longer, becoming comparable with the laths size. The formation of LTUB at 400°C and HTUB at 450°C proceeds through the mechanism of lateral growth with a layer-by-layer accretion of laths separated by carbides. As a result, the microstructure is composed of the packets of wide laths with a longitudinal layout of coarse cementite carbides. These laths alternate with narrower laths having carbides precipitated inside the lath body at an angle of 50-60 degrees to the longitudinal axis. High temperature upper bainite is characterized by a decreased density of lattice defect in α-phase laths.

Analyzing the data obtained, the beneficial microstructure and substructures of steels 75Mn and 75Mn2Si, having "Lower Massive bainite" (250°C) should be highlighted. This structure is characterized by the presence of highly dispersed carbides precipitated inside ferrite laths while laths boundaries are free of carbides, which is beneficial for mechanical properties. Dispersed carbides inside the laths result in the composite strengthening of ferrite laths while the retained austenite films ensure increased ductility of LMB.

In contrast to steels 75Mn and 75Mn2Si "Lower Massive Bainite" formed in steel 75MnCrSiMo is characterized by the initial stage of carbide precipitation along the laths boundaries, which could negatively affect the fracture toughness. In addition, LMB formation in steel 75MnCrSiMo was accompanied with dispersion strengthening of retained austenite, which can reduce austenite ability for stress relaxation through micro-strain deformation. It can be concluded that the presence of strong carbide-forming elements (Cr, Mo) in steel 75MnCrSiMo stimulates the precipitation of carbides from austenite during bainite transformation. This chromium behaviour was described earlier in [39]. As found in [40], that even a small increase in the chromium concentration intensifies the formation of a carbide network along the austenite grain boundaries when rolled products made of bearing steels are cooled in the air; this is explained by the reduction in the solubility of carbon in austenite under the chromium effect.

Chromium reduces the carbon activity in austenite, thus contributing to a decrease in the chemical energy of the system [41]. It is known that at low temperature holding the formation of carbides does not involve carbide-forming elements because of their lower diffusivity, which requires significant thermal activation [46]. Consequently, when carbides are precipitating, an increase in the chromium concentration in the austenite adjacent to the carbides should be observed, which, in turn, leads to an additional decrease in carbon activity, which is an energetically beneficial process. At the same time, steel 75MnCrSiMo contains an elevated (relative to other studied steels) content of silicon. Unlike chromium, silicon inhibits the precipitation of carbides [42, 43]. Silicon is not dissolved in carbide, thus when carbide grows, the silicon is pushed away from the carbide into the austenite matrix causing the significant increase in carbon activity in austenite [39, 41]. This makes carbide precipitation to be energetically unfavourable.

As follows from TEM observation, the superposition of the effects of silicon and chromium results in the precipitation of carbides from austenite in steel 75MnCrSiMo during LMB formation (which differs from steels 75Mn and 75Mn2Si behaviour). To confirm this assumption, the change in the free chemical energy of austenite under the influence of alloying elements was calculated using the approach explicated in [41]:

$$\Delta G^{\gamma} = \mathrm{x}_c \cdot R \cdot T \cdot \ln a_C^{\gamma} + (1 - x_c) \cdot R \cdot T \cdot \ln a_{Fe}^{\gamma} \tag{6.1}$$

where T is absolute temperature, x_c is the molar fraction of carbon in austenite, and a_C^{γ} and a_{Fe}^{γ} are the activities of carbon and iron in austenite, respectively.

Carbon activity in austenite in alloyed steel was found as [41]:

$$a_c^{\gamma} = \gamma_c \cdot x_C \cdot f^{ae_1} \cdot f^{ae_2} \cdot \ldots \cdot f^{ae_i}, \tag{6.2}$$

where γ_c is carbon activity in unalloyed austenite, f^{ae_i} is the coefficient which reflects the effect of *i*-th alloying element on carbon activity.

To determine γ_c coefficients the following expression was used [41]:

$$\gamma_c = x_c / (1 - 5x_c) \tag{6.3}$$

The iron activity in austenite was calculated as [41]:

$$lg\, a_{Fe}^{\gamma} = 0{,}25\, lg\left(\frac{1-5x_c}{1-x_c}\right) - \frac{158}{T}\left(\frac{x_c^2}{(1-x_c)^2}\right) \tag{6.4}$$

The coefficients f^{ae_i} were determined in accordance with [41]:

$$\ln f_c^{Si} = 10{,}35 \cdot x_{Si} \tag{6.5}$$

$$f_c^{Mm} = -(6500/T) + 1.45 \tag{6.6}$$

$$f_c^{Cr} = -(21900/T) + 6.3 \tag{6.7}$$

$$f_c^{Mo} = -(37600/T) + 16.9 \tag{6.8}$$

where x_{Si} is the Si mole fraction in austenite.

The calculation results are shown in Figure 6.7. As follows from this figure, in steel 75MnCrSiMo, the effect of chromium on the free energy prevails over the influence of silicon, which is expressed in the minimum (from all three steels) values of ΔG at different holding temperatures. Consequently, since carbide formation is accompanied by the accumulation of chromium in the matrix, then carbide precipitation from austenite under bainite transformation might proceed more intensively in steel 75MnCrSiMo than that in steels 75Mn and 75Mn2Si. This is confirmed by experimental results shown above.

The analysis of the austenite transformation in the studied steels was performed using Johnson–Mehl–Avrami equation (1.1) with a double logarithm which yielded its linear form:

$$ln\,[-ln\,(1\text{-}f)] = lnK + n ln\,\tau, \qquad (6.9)$$

where f is the volume fraction of transformed austenite; τ is the transformation duration; K is a coefficient; n is the kinetic index.

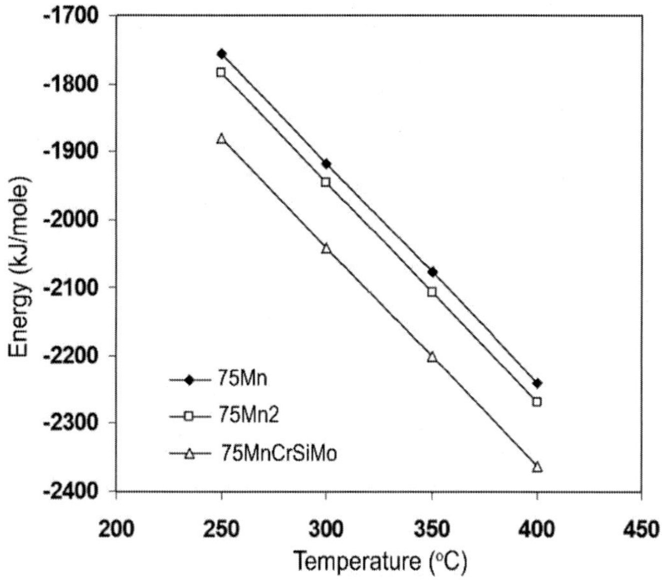

Figure 6.7. The effect of temperature on austenite free energy in studied steels.

The kinetic curves of austenite transformation in steels 75Mn, 75Mn2Si and 75MnCrSiMo (Figure 4.1, 4.7, 4.11), transformed by the double logarithmization, are presented in Figure 6.8 as lines. Most of the lines consist of two straight sections with a different inclination to the abscissa axis, therefore differing in the value of the index n in Johnson–Mehl–Avrami equation. The data on n values for different steels are summarized in Table 6.1.

The kinetic transformation curves for steels 75Mn and 75Mn2Si related to the pearlitic (i.e., diffusional) transformation interval (450-650°C and

500-650°C, respectively) have in the first section the values of $n$in the range of 3.6-5.0, which is close to n = 4. This n value, according to Cahn [1], corresponds to a constant or increasing rate of α-phase nucleation. In the second section, the value of n is in the interval of 1.4-2.2, which corresponds to the nucleation on the edges of the grains after the sites of nucleation have been exhausted. In steel 75MnCrSiMo in the temperature range of pearlitic transformation the n value at the first stage is about 4 only at 500°C.

At higher holding temperatures, the values of the n index for steel 75MnCrSiMo are increased to 7.59-9.34, which indicates an increasing rate of transformation at its first stage. In the second stage, n is 3.1-6.4; this corresponds either to an increase in the transformation rate (for n>4) or to the exhaustion of nucleation sites (for n = 3).

In the range of bainitic transformation in steels 75Mn and 75MnCrSiMo, the index n values for the first and for the second stages are 2.26-2.58 and 1.49-1.54, respectively. Assuming (based on existing concepts [44]) that the bainitic transformation is controlled by carbon diffusion and proceeds through carbide precipitation, then the given values of n_1 can be addressed to the formation of carbide nuclei of different shapes (at constant nucleation rate), and n_2 – to the growth of carbide particles at zero nucleation rate.

The specific feature of steel 75Mn2Si is that in the lower temperature range (at 250-300°C), after the first and the second transformation stages (with n_1 = 2.21-2.22 and n_2= 0.9-1.5 respectively), the third stage (n_3= 2.9-3.1) with an increased transformation rate appears. At higher holding temperatures (350 – 400°C), two transformation stages are observed; however, unlike steels 75Mn and 75MnCrSiMo, at the second stage, the nucleation rate slightly increases to n_2 = 2.9-3.1.

These features indicate a more complex mechanism of the bainitic transformation in steel 75Mn2Si. The probable explanation is an increased content of manganese in the steel, which significantly reduces the martensitic point Ms in austenite. This requires more significant carbon partitioning in austenite to create much depleted austenite areas with an increased Ms point where displacive transformation is possible. Obviously, this has the result that the highest rate of the transformation shifts to the second stage or even to the third stage of the transformation, i.e., by the time the carbon partitioning is complete. This assumption is confirmed by the fact that the three-stage transformation is characteristic precisely for the lowest range of holding temperatures (250-300°C).

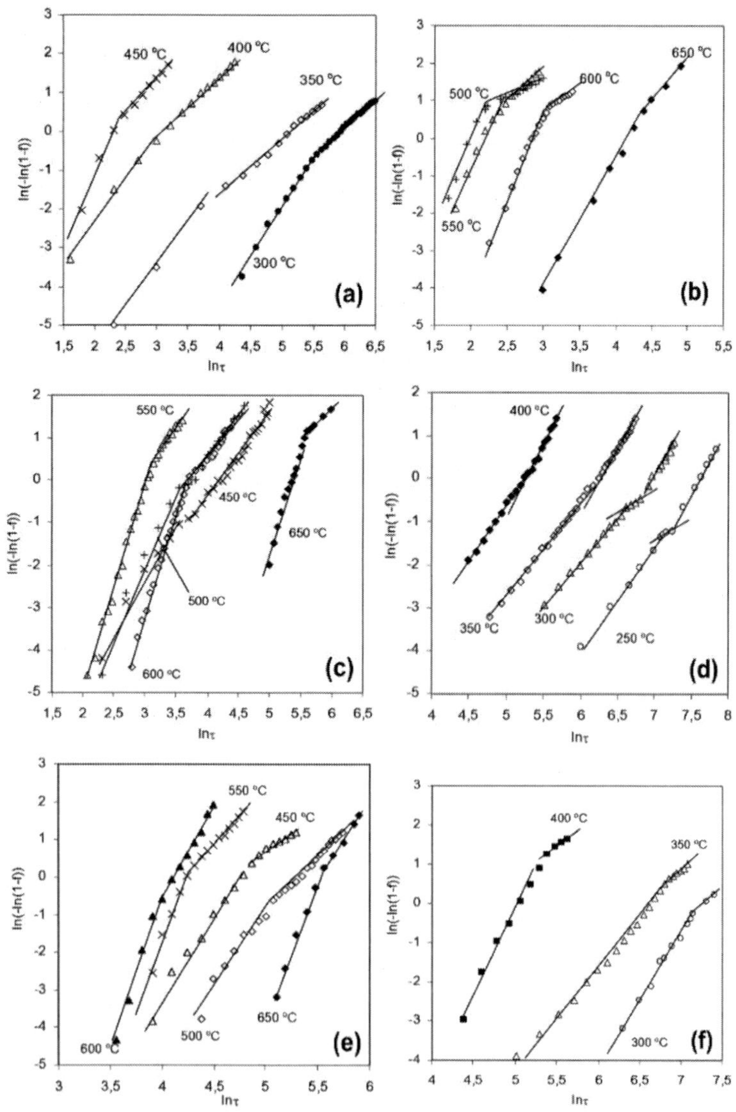

Figure 6.8. Kinetics curves according to Equation (6.9) for steels: (a, b) 75Mn, (c, d) 75Mn2Si, (e, f) 75MnCrSiMo.

In this case, the transformation actually goes through the suspension in the second stage (zero nucleation rate) before the acceleration in the third stage. It can be assumed that this is due to the need to carry out preparatory processes (carbon partitioning) under conditions of minimal diffusivity of carbon atoms.

Table 6.1. The values of index *n* in Johnson–Mehl–Avrami equation

Temperature (°C)	75Mn		75Mn2Si			75MnCrSiMo	
	n_1	n_2	n_1	n_2	n_3	n_1	n_2
250	-	-	2.22	0.94	3.08	-	-
300	2.39	1.51	2.21	1.53	2.93	2.58	1.42
350	2.26	1.52	2.01	2.86	-	2.47	1.49
400	2.22	1.54	2.63	3.05	-	4.08	1.62
450	3.93	1.61	2.46	1.94	-	4.07	1.67
500	4,44	0.94	3.76	2.06	-	4.05	3.10
550	4.70	1.82	4.85	2.12	-	7.59	3.09
600	4.51	1.50	4.71	1.77	-	9.34	4.98
650	3.58	2.20	5.04	1.43	-	8.59	6.38

CONCLUSION

1. The kinetics and mechanism of austenite transformation in steels 75Mn, 75Mn2Si and 75MnCrSiMo under isothermal holding and with continuous cooling are studied and discussed; the corresponding TTT and CCT diagrams are constructed and presented in the paper.
2. It has been found that under isothermal conditions steel 75Mn is characterized by one kinetic maximum of austenite transformation corresponding to 400-500°C with a minimal incubation period of 4 sec for 450-500°C. Steel 75Mn has a pearlitic type of hardenability, and the critical quenching rate for this steel is found to be about 30 K/sec.

   Austenite transformation in steel 75Mn2Si under isothermal conditions is characterized by one kinetics maximum at 450-550°C, and the minimum incubation period is detected as 7 sec at 550°C.

Steel 75Mn2Si has a bainitic type of hardenability; its critical quenching rate is about 10 K/s.

Austenite transformation in steel 75MnCrSiMo under isothermal conditions is characterized by two kinetic maxima which are 450°C (the incubation period is 50 sec) and 600°C (the incubation period is 38 sec). Steel 75MnCrSiMo has a bainitic type of hardenability; its critical quenching rate is 13.2 K/s.

3. With continuous cooling the different combinations of microstructure could be obtained in the studied steels depending on cooling rate. These are for steel 75Mn: martensite, martensite/pearlite, pearlite; for steel 75Mn2Si: martensite, martensite/bainite, martensite/bainite/pearlite, pearlite; and for steel 75MnCrSiMo: martensite, martensite/bainite, bainite/pearlite, pearlite.
4. The characteristic features of bainitic structures observed in steels 75Mn, 75Mn2Si and 75MnCrSiMo are revealed and discussed. These structures could be classified morphologically as: (a) "Lower Needle Bainite", formed at 200ºC in steels 75Mn2Si and 75MnCrSiMo, (b) "Lower Massive Bainite" formed at 250-300°C in steels 75Mn, 75Mn2Si, and 75MnCrSiMo, (c) "Low Temperature Upper Bainite" formed at 300-350°C in steels 75Mn, 75Mn2Si, and 75MnCrSiMo, (d) "High Temperature Upper Bainite" formed at 400-450°C in steels 75Mn2Si, and 75MnCrSiMo.
5. The features of bainite substructure in steels 75Mn, 75Mn2Si, and 75MnCrSiMo are revealed. It is established that an increase in the temperature of isothermal holding is accompanied by: (a) thickening of α-phase laths, (b) enlargement of carbide particles with decrease in their number, (c) decreasing in α-phase laths and increasing lattice defects in carbide particles.
6. The peculiarities of carbides distribution inside α-phase laths of various morphological bainite types are established. It is shown that in "Lower Needle Bainite" and in "Lower Massive Bainite" carbides are located inside the laths at an angle of 55-70 degrees to its long axis. In this case, the laths boundaries (except in steel 75MnCrSiMo) are free of carbides. In "Low Temperature Upper

Bainite" the carbides are precipitated inside the laths as well as between the laths. In "High Temperature Upper Bainite" the carbides are located mostly along the laths boundaries.

7. The "Lower Massive Bainite" in steel 75Mn2Si is characterized by the presence of austenitic interlayers (films) between α-phase laths and by the presence of fibrous carbides in retained austenite.
8. For the "Lower Massive Bainite" in steel 75MnCrSiMo, the precipitation of carbide along laths boundaries and the precipitation of dispersed carbides in retained austenite are typical, which could negatively affect the fracture toughness of steel 75MnCrSiMo.
9. Analysis of the experimental data basing on Johnson–Mehl–Avrami equation and Cahn's theory revealed that austenite transformation in steels 75Mn and 75Mn2Si is characterized by a two-stage process, consisting of the nucleation stage with a constant/ increasing rate followed by a nucleation stage on the grain edges after exhaustion of the nucleation sites. The austenite transformation in steel 75MnCrSiMo features a three-stage transformation (at 250-300°C), with a decreasing transformation rate at the second stage followed by a further increasing rate in the third stage. This behaviour may be caused by a longer carbon redistribution which is necessary to continue the bainitic transformation in this steel.

ACKNOWLEDGMENTS

The authors acknowledge the financial support from Ministry of Education and Science of Ukraine (project 0117U002270).

REFERENCES

[1] Cahn, R.W. & Haasen, P. (1996). *Physical Metallurgy*. North Holland.

[2] Morozov, O.P. & Schastlivtsev, V.M. (1988). "Low-Temperature Pearlite in Plane High-Carbon Steels". *Fiz. Met. Metalloved.*, *66*(5), 910-919.

[3] Yun-Chul, J., Hiroyuki, O., Kiyomichi, N. & Yasuya, O. (1996). "Isothermal Decomposition Processes of Austenite in Fe-Cr-C Alloy Steels". *Mater. Transaction, JIM*, *37*(4), 676-683.

[4] Weinberg, Michael C. (1987). "A Test of the Johnson-Mehl-Avrami Equation". *Journal of Crystal Growth*, *82*(4), 779-780

[5] Gur, C.H. & Pan, J. (2009). *Handbook of Thermal Process Modelling Steels*. Taylor & Francis Group.

[6] Hillert, M. "Discussion of "A Commentary on Reaction Kinetics in Processes of Nucleation and Growth". *Metall. Mater. Trans. A*, *42A*, 3241

[7] Schastlivtsev, V.M., Yakovleva, I.L. & Mirzaev, D.A. (1994) "Structural Transformations in Pearlite under Heating. I. The Origin of Transformation Hardening and the Recrystallization of Ferrite", *Fiz. Met. Metalloved.*, *77*(4), 138-147.

[8] Verhoeven, J.D. & Gibson, E.D. (1998). "The Divorced Eutectoid Transformation in Steel". *Metall. Mater. Trans. A*, *29*(4), 1181-1189.

[9] Han, K., Smith, G.D. W. & Edmonds, D.V. (1995). "Pearlite Phase Transformation in Si and V Steel". *Metall. Mater. Trans. A*, *26A*, 1617–1631.

[10] Elwazri, A.M., Wanjara, P. & Yue, S. (2006). "Empirical Modelling of the Isothermal Transformation of Pearlite in Hypereutectoid Steel". *Mater. Sci. Technol.*, *22*(5), 542-546(5).

[11] Yudin, Yu.V. & Farber, V.M. (2001). "Decomposition Kinetics of Under-Cooled Austenite of Alloy Steels Within the Pearlitic Range". *Metal Sci. Heat Treat.*, *2*, 3-9.

[12] Yakovleva, I.L., Kar'kina, L.E., Zubkova, T.A. & Tabatchikova, T.I. (2011). "Effect of Cold Plastic Deformation on the Structure of Granular Pearlite in Carbon Steels". *Phys. Metals Metallogr.*, *112*, 101-108.

[13] Parusov, V.V. Dolzhenkov, I.I. & Sukhomlin, V.I. (1985). "Transformation of Austenite into Granular Pearlite in Carbon and Alloy Steels". *Metal Sci. Heat Treat.*, *27* (6), 402–407.

[14] Popova, L.E. & Popov, A.A. (1991). *Diagrams of Austenite Transformation in Steels and β-solution in Titanium Alloys*. Moscow: Metallurgiya (in Russian).

[15] Novikov, I.I. (1986). *Theory of Heat Treatment of Metals*. Moscow: Metallurgiya. (in Russian).

[16] Aleksandrov, L.N. & Lyubov, B.Ya. (1963). "Rost Kristallov Beynita [Growth of Crystals of Bainite]". *Doklady AN SSSR - Reports of AS USSR,151*(3), 552-556. (in Russian).

[17] Langer, J.S. (1971). "Theory of Spinodal Decomposition in Alloys". *Annals of Physics,65*(1), 53-86.

[18] Schastlivtsev, V.M. (2005). "New Concepts of the Nature of Bainitic Transformations in Steels". *Met. Sci. Heat Treat.,47*(7-8), 276-281.

[19] Teplukhin, G.N. & Danilov, A.N. (1983). "Varieties of Upper Bainite in Steels of the Pearlitic Class". *Metal Sci. Heat Treat.*, *25*(10), 785–788.

[20] Teplukhin, G.N. (1993). "Various Features of the Transformations in Pearlitic Steels". *Metal Sci. Heat Treat.*, *7*, 23-26.

[21] Morozov, O.P., Schastlivtsev, V.M. & Yakovleva, I.L. (1990). "Upper and Lower Bainite in Plane Eutectoid Steel". *Fiz. Met. Metalloved.,2*, 150-159.

[22] Krauss, G. (2015). *Steels: Processing, Structure, and Performance*, ASM International.

[23] Cheprasov, D.P., Svishchenko, V.V., Kozlov, E.V. & Ivanaiskii, A.A. (2006). "Phase Composition and Fine Structure of Granular Bainite in Low-Carbon Low-Alloy Steel". *Metal Sci. Heat Treat.*, *48*(5-6), 187–192.

[24] Gudremon, E. (1959). *Special Steels*. Moscow: Metallurgizdat (in Russian).

[25] Mes'kin, V.S. (1964). *Basements for Steel Alloying*. Moscow: Metallurgiya (in Russian).

[26] Pickering, F.B. (1982). *Physical Metallurgy and the Design of Steels*. Moscow: Metallurgiya. (in Russian).

[27] Perez-Unzueta, A. J. & Beynon, J.H. (1993). "Microstructure and Wear Resistance of Pearlitic Rail Steels". *Wear*, 162–164, 173-182.

[28] Nesterov, D.K., Razin'kova, N.N., Chernyakova, L.E. & Kisil, B.S. (1991). "Structure and Mechanical Properties of Rails Type R65 Bulk-Hardened in Oil". *Metal Sci. Heat Treat.*, *33*(4), 331–336.

[29] Lempitskyi, V.V., Kazarnovskyi, D.S. & Gubert, S.B. (1972). *Productiong and Heat Treatment of Rails*. Moscow: Metallurgiya. (in Russian).

[30] Nesterov, D.K., Razin'kova, N.N. & Vorozhishchev, V.I. (1988). "Improving the Heat Treatment Technology of Rails", *Stal'*, *1*, 75–79.

[31] Razin'kova, N.N., Kisil', B.S. & Levchenko, N.F. (1986). "Refinement of the Parameters of Bulk-Hardening of Rails Made of Steels Deoxidized by a Complex Alloy". *Stal'*, *7*, 83–85.

[32] Nesterov, D.K., Sapozhkov, V.E., Levchenko, N.F. & Dubrov, V.A. (1990). "Heat Treatment of Rail Steel Using Induction Heating". *Metal Sci. Heat Treat.*, *32*(8), 589-594.

[33] Nesterov, D.K., Levchenko, N.F., Sapozhkov, V.E. & Bashkatova, Zh.B. (1989). "Steel for Rails of Increased Strength". *Stal'*, *12*, 67–70.

[34] Caballero, F.G., Bhadeshia, H.K.D.H., Mawella, K.J.A., Jones, D.G. & Brown, P. (2002). "Very Strong Low-Temperature Bainite", *Mater. Sci. Technol.*, *18*, 279-284.

[35] Singh, U.P., Roy, B., Jha, S. & Bhatacharrayya, S.K. (2001). "Microstructure and Mechanical Properties of as Rolled High Strength Bainitic Rail Steels". *Mater. Sci. Technol.*, *17*, 33-38.

[36] Lowrinowicz, Z. & Barbacki, A. (2002). "Features of Bainite Transformation in Steels". *Advances in Material Science*,*1*(2), 5-27.

[37] Pacyna, J. (2008). "The Microstructure and of the New Bainitic Rail Steels". *Journal of Achievements in Materials and Manufacturing Engineering*, *28*, 19-22.

[38] Kuz'min, S.O., Efremenko, V.G., Chabak, Yu. G. & Tsvetkova, E. V.(2013). "Morphologic Features of Bainite in High-Carbon Complex-Alloyed Steel". *Metallofiz. Noveishie Tekhnol.*, *35*(9), 1271–1282.

[39] Kurdiumov, G.V., Utevskii, L.M. & Entin, R.I. (1977). *Transformations in the iron and steel*. Moscow: Nauka (in Russian).

[40] Yatsenko, Yu. V. & Spektor, Ya.I. (1984). "Carbide Network Formation in High-Carbon Steels". *Stal'*, *10*, 75-76.

[41] Mogutnov, B.M., Tomilin, I.A. & Shvartzman, L.A. (1984). *Thermodynamics of Iron Alloys.* Moscow: Metallurgiya. (in Russian).

[42] Mašek, B., Jirková, H., Hauserova, D., Kučerová, L. & Klauberová, D. (2010). "The Effect of Mn and Si on the Properties of Advanced High Strength Steels Processed by Quenching and Partitioning". *Materials Science Forum*, *654*,94-97.

[43] Garcia-Mateo, C., Caballero, F.G., Sourmail, T., Kuntz, M., Cornide, J.& Elvira, R. (2012). "Tensile Behaviour of a Nanocrystalline Bainitic Steel Containing 3wt% Silicon". *Materials Science and Engineering: A*, *549*, 185-192.

[44] Bhadeshia, H.K.D.H. (2001). *Bainite in Steels: Transformations, Microstructure and Properties*. IOM Communications.

BIOGRAPHICAL SKETCH

Efremenko Vasily

Affiliation: Priazovskyi State Technical University

Education: Engineer. Specialty - Metal Science and Heat Treatment of Metals, graduated from Priazovskyi State Technical University (PSTU), Mariupol

Research and Professional Experience:

- Candidate of Science (analogues PhD). (1989). Specialty - Metal Science and Heat Treatment of Metals
- Doctor of Science (2006). Specialty - Metal Science and Heat Treatment of Metals.

- From 1988 up to now has participated in 30 RD/innovative projects funded by industrial enterprises of Ukraine. Since 2008 has participated as a Project Leader (totally in 10 RD/innovative projects).

Professional Appointments: Since 1990 up to now working at PSTU (from Assistant Professor to Professor, Head of Physics Department).

- Honorable Diploma from Ministry of Science and Education of Ukraine
- Honorable Diploma from Government of Ukraine

Publications from the Last 3 Years:

[1] Efremenko, V.G., Wu, K. M., Chabak, Yu. G., Shimizu, K., Isayev, O.B. & Kudin, V.V. (2018). Alternative Heat Treatments for Complex-Alloyed High-Cr Cast Iron Before Machining. *Metallurgical and Materials Transactions A.*, Vol. *49*, Is. 8, p. 3430–3440.

[2] Chabak, Yu. G., Efremenko, V.G., Shimizu, K., Lekatou, A., Pastukhova, T.V., Azarkhov, A. Yu. & Zurnadzhy, V.I. (2018). Comparative Analysis of the Microstructural Features of 28 wt.% Cr Cast Iron Fabricated by Pulsed Plasma Deposition and Conventional Casting. *Journal of Materials Engineering and Performance.*, Vol. *27*, Number 2,p. 379-388.

[3] Efremenko, V., Shimizu, K., Chabak, Yu., Brykov, M., Kusumoto, K. & Efremenko, A. (2018).Three-body abrasive wear behaviour of metastable spheroidal carbide cast irons with different chromium contents. *International Journal of Materials Research.*, Vol. *109*, No. 2,p. 147-156.

[4] Efremenko, V.G., Zurnadzhi, V.I., Chabak, Y. G., Tsvetkova, O.V. & Dzherenova, A. V. (2017). Application of the Q-n-P-Treatment for Increasing the Wear Resistance of Low-Alloy Steel with 0.75% C, *Material Science.*, Vol. *53*,p. 67-75.

[5] Chabak, Yu. G., Pastukhova, T.V., Efremenko, V.G., Shimizu, K., Cheiliakh, A.P. & Zurnadzhy, V.I. (2017). Features of Formation of Microstructure, Elemental and Phase Compositions, and Properties of the 1.7%C–14%Cr–3%Mn–3%Si–1%Ni–0.6%V–1.2%B Steel under

Casting and Pulsed Plasma Deposition. *Metallofizika i Noveishie Tekhnologii.*, Vol. *39*, p. 491-505.

[6] Efremenko, V.G., Chabak, Yu. G., Karantzalis, A.E., Lekatou, A., Vakulenko, I.A., Mazur, V.A. & Fedun, V.I. (2017). Plasma Case Hardening of Wear-Resistant High-Chromium Cast Iron. *Strength of Materials.*, Vol. *49*, Is. 3,p.446–452.

[7] Zhang, Y., Shimizu, K., Yaer, X., Kusumoto, K. & Efremenko, V.G. (2017). Erosive wear performance of heat treated multi-component cast iron containing Cr, V, Mn and Ni eroded by alumina spheres at elevated temperatures, *Wear.*, vol. 390-391,p. 135-145.

[8] Chabak, Yu. G., Fedun, V.I., Pastukhova, T.V., Zurnadzhy, V.I., Berezhnyy, S.P. & Efremenko, V.G. (2017). Modification of steel surface by pulsed plasma heating. *Problems of Atomic Science and Technology.*, Vol. *110*, No 4.,p. 97-102.

[9] Efremenko, V.G., Chabak, Yu. G., Shimizu, K., Lekatou, A.G., Zurnadzhy, V.I., Karantzalis, A.E., Halfa, H., Mazur, V.A. & Efremenko, B.V. (2017). Structure refinement of high-Cr cast iron by plasma surface melting and post-heat treatment. *Materials & Design.*, Vol. *126*,p. 278–290.

[10] Efremenko, V.G., Shimizu, K., Pastukhova, T.V., Chabak, Yu. G., Kusumoto, K. & Efremenko, A.V. (2017). Effect of bulk heat treatment and plasma surface hardening on the microstructure and erosion wear resistance of complex-alloyed cast irons with spheroidal vanadium carbides. *Journal of Friction and Wear.*, Vol. *38* (1),p. 58-64.

[11] Efremenko, V.G., Shimizu, K., Pastukhova, T.V., Chabak, Yu. G., Kusumoto, K. & Efremenko, A.V. (2017). Wear mechanism and chemical composition optimization of complex-alloyed cast iron with spheroidal vanadium carbide under conditions of abrasive erosion. *Journal of Friction and Wear.*, Vol. *38* (2),p. 92-97.

[12] Efremenko, V.G., Chabak, Yu. G., Lekatou, A., Karantzalis, A.E., Shimizu, K., Fedun, V.I., Azarkhov, A. Yu. & Efremenko, A.V. (2016). Pulsed plasma deposition of Fe-C-Cr-W coating on high-Cr-cast iron: Effect of layered morphology and heat treatment on the

microstructure and hardness. *Surface and Coatings Technology.*, Vol. *304*, p. 293–305.

[13] Chabak, Y.G., Fedun, V.I., Shimizu, K., Efremenko, V.G. & Zurnadzhy, V.I. (2016). *Phase-structural composition* of *coating obtained by pulsed plasma treatment using eroded cathode* of *T1 high speed steel. Problems of Atomic Science and Technology. Series Plasma Electronics and New Acceleration Methods.*, № 4.,p. 100-106.

[14] Efremenko, V.G., Shimizu, K., Cheiliakh, A.P., Pastukhova, T.V., Chabak, Yu. G. & Kusumoto, K. (2016). Abrasive resistance of metastable V–Cr–Mn–Ni spheroidal carbide cast irons using the factorial design method. *International Journal of Minerals, Metallurgy, and Materials.*, Vol. *23*(6),p. 645-657.

[15] Efremenko, V.G., Chabak, Yu. G., Lekatou, A., Karantzalis, A. E. & Efremenko, A.V. (2016). High-temperature oxidation and decarburization of 14.55 wt pct Cr-cast Iron in dry air atmosphere. *Metallurgical and Materials Transactions A.*, Vol. *47A* (2), p. 1529-1543.

INDEX

#

A

B

E

F

G

H

I

J

K

L

M

Q

R

S

V

W

Y

Z

Advances in Materials Science Research Volume 37

Editor: Maryann C. Wythers

Series: Advances in Materials Science Research

Book Description: Advances in Materials Science Research. Volume 37 begins by discussing the main properties of poly (methyl methacrylate) bone cements used for orthopaedic surgery.

Hardcover ISBN: 978-1-53615-038-4
Retail Price: $250

Advances in Materials Science Research Volume 36

Editor: Maryann C. Wythers

Series: Advances in Materials Science Research

Book Description: In the preliminary study included in Advances in Materials Science Research. Volume 36, a novel fiberglass catalyst with highly dispersed 1-2 nm Pd clusters confined in the bulk of fiberglass at a depth of 10 nm was proposed.

Hardcover ISBN: 978-1-53614-848-0
Retail Price: $250